ROYAL COLLEGE OF ART SCHOOLS TECHNOLOGY PROJECT

D&T CHALLENGES

DESIGN & TECHNOLOGY 11-14

SCHOOLS
In association with the
CHANNEL 4 SCHOOLS series
Real Life Design

A CTC Trust programme sponsored and
supported by the Royal College of Art,
the Esmée Fairbairn Trust and the
Department for Education and Employment

Hodder & Stoughton

A MEMBER OF THE HODDER HEADLINE GROUP

Acknowledgements

The publishers would like to thank the following:

Danny Jenkins of fab 4 studio for the cover illustration; Lynda King for the cover and book design; John Urling Clark for picture research; Keith Howard, Colin King, Sally Launder, Frank Nichols, Martin Orme, Gary Rees, Martin Salisbury, Sara Silcock and Jacqui Thomas of Linda Rogers Associates for the illustrations; Koichi Saikyo and Ela Cieslak of the Royal College of Art.

We are also grateful to the following companies, agencies and individuals who have given permission to reproduce photos and artwork in this book. Every effort has been made to trace and acknowledge ownership of copyright. The publishers will be glad to make suitable arrangements with any copyright holder whom it has not been possible to contact.

J. Allen Cash (11 top right, 17 top centre top and middle, 24/25, 32 left and first right, 33, 50 right centre, 72 bottom, 78 fourth right, 84 centre); Allied Signal (122); Anthony Blake Photo Library (78 left, third right, 79); Anthony Blake Photo Library © Merehurst (78 second right); Ron Arad Associates (17 top right); Artidee, Alkmar from *The Eye Beguiled* by Bruno Ernst, Taschen (9 bottom centre and right); Kathie Atkinson/Oxford Scientific Films (14 bottom centre); Alex Bartel/Science Photo Library(50 left, 51); John Birdsall (17 top left, top centre bottom, 18/19, 22 top and all bottom, 23 centre and bottom, 24 left, 26, 29, 30 all, 31 both, 40, 44 left and right top, 47 both, 49 top, 51 right bottom, 52 bottom, 55, 59 middle, 86 all bottom, 88 all bottom, 94 both, 112 all, 113 all, 119 both, 120 bottom right, 121 both, 125 all, 126 all); Alan Booth (94 bottom); BP (88 both top left); Bridgeman Art Library (10 top, 77 centre); British Rail (88 right); British Standards Institute (42 top); Brooke Calverly (8), Camden Food and Drink Research (116 top right); Collections/John and Eliza Forder (38 left); Collections/Shout (38 right centre); Collections/Brian Shuel (120 left top right, left bottom, top right); Bruce Davidson/Oxford Scientific Films (12 first right); Chris Davies (97 both, 100, 108 all, 109 both); De Roma Ice Cream (83); Jean Larcher/Dover Books (10 bottom), Ivan Ripley (9 top upper centre, top right), Lawrence Whistler (9 top left), (9 top lower centre) all from *Visual Magic* by Dr David Thomson, Dial Books; © 1995 M.C.Escher/Cordon Art, Baarn, Holland. All rights reserved (6/7 both, 9 bottom left, 11 top left); Eden Vale (90 right); Eye Catchers Press/BT Corporate Pictures (52 top); Eye Ubiquitous (76 top left); Andy Faris/Robert Harding Picture Library (44 centre right); M.P.L. Fogden/Oxford Scientific Films (12 left); Fraser Photographers/Crosrol (23 both top); Carlos Freire/Hutchinson Library (70 centre); Lucy Goffen (Eye Ubiquitous (76 top left); Maggie Grey (63, 70 top); Joaquin Gutierrez Acha/Oxford Scientific Films (14 bottom top); Halfords (57 top, 60 both); Jeremy Hartley/Oxfam (73); Heinz (28 bottom right); Andrew Hirniak (66 top, bottom both left and centre); Kit Houghton Photography (59 top); Jacqui Hurst (70 bottom, 75 left, 77 bottom); IKEA (98); B. & G. Ingram-Monk (65 bottom); J.Jackson/Robert Harding Picture Library (45); David Kampfner (32 third right); Emma Lee/Life File (11 bottom, 48 right, 52 right and detail, 67 bottom left); R.Ian Lloyd/Hutchinson Library (71); C.Lucas/Oxfam (76 right); S.Kay/Life File (32 second right); London Transport (88 lower top left); Michael Macintyre/Hutchinson Library (44 right bottom, 64 fourth right); Robert McBain/Maynard Leigh Associates (87 right); John Mead/Science Photo Library (50 right top, 52 top left) by courtesy of Mercury Records Ltd (11 bottom right); Mary Moran (64 left, second and third right); John Myerson (58 bottom); Louise Oldroyd/Life File (59 bottom); Oxfam (77 top); Petzl (42, 98 top); Roddy Paine (12 fifth right, 14 top right, 16 all, 17 bottom, 32 fifth right, 34, 37, 103 top); Phillips 39); Tanya Piejus (11 bottom top centre), 72 top left); Peter and Mary Plage/Oxford Scientific Films (12 second right); Dr Morley Read/Science Photo Library (14 bottom); Robert Harding Picture Library Ltd (15 top right, 65 top); Royal Mail (11 bottom top right); Brian Russell (56 both, 57 top, 114 all, 124 all); David Scharf/Science Photo Library (13); Krupaker Senani/Oxford Scientific Films (12 fourth right); Southampton City Art Gallery (66 bottom right); students of Partner Schools (20 top, 22 top, 35); Trip/M.Barlow (103 bottom); Trip/G.Howe (15 bottom, 86 top, 87 left, 95 both, 96, 102 both, 118, 123); Trip/H.Rogers (15 top left, top centre, 18 left and centre, right top and bottom, 46 left, 62, 67 top left, 82, 84 left, 84/85, 120 left top left); Trip/B.Turner (72 top); Kalvin Turner (115 both); Twentieth Century Fox (12 centre); John Urling Clark (12 third right, 46 top right, 67 bottom left); Andrew Watson (fourth right); Andy Williams/Robert Harding Picture Library (48 left); Williams Photography Ltd/Heinz (27, 28 top and bottom left); Val & Alan Wilkinson/Hutchinson Library (76 bottom); Carol Wills/Oxfam (75 bottom); World Crafts (75 top); George Wright/Hincliffe and Barber (49 both bottom).

Cataloguing in Publication Data is available from the British Library.

ISBN 0 340 63928 8

First published 1996
Impression number 10 9 8 7 6 5 4 3 2 1
Year   2000 1999 1998 1997 1996

Copyright © 1996 City Technology College Trust, Schools Technology Project works under the auspices of the Royal College of Art.

All rights reserved. No part of this publication may be reproduced or transmitted in any form or by any means, electronic or mechanical, including photocopy, recording or any information storage and retrieval system, without permission in writing from the publisher or under licence from the Copyright Licensing Agency Limited. Further details of such licences (for reprographic reproduction) may be obtained from the Copyright Licensing Agency Limited, of 90 Tottenham Court Road, London W1P 9HE.

Typeset by Wearset, Boldon, Tyne and Wear.
Printed in Great Britain for Hodder and Stoughton Educational, a division of Hodder Headline Plc, 338 Euston Road, London NW1 3BH by Cambridge University Press, Cambridge.

ROYAL COLLEGE OF ART SCHOOLS TECHNOLOGY PROJECT

The Royal College of Art Schools Technology Project is a three year programme which started in September 1993. It is designed to raise the sights and cater for the curricular needs of technology teachers in secondary schools. The project is funded by the Esmée Fairbairn Charitable Trust, Cable and Wireless plc and the Department for Education and Employment.

The underlying purpose is to improve the quality of technology education throughout secondary schools in England.

The Project's central team is based at the Royal College of Art, supporting developments in 13 selected secondary schools with Teacher Fellows on part-time secondment. They are developing a comprehensive design and technology course for students aged 11 to 19. The whole team is working closely with business organisations and major industrial companies to ensure a curriculum which is relevant to the worlds of business and commerce.

This student book forms part of the course developed to deliver the requirements of, and to enhance, the National Curriculum and post-14 work, particularly for GNVQ.

Royal College of Art Schools Technology Project 1995 Writing Team

Teacher Fellows and Partner Schools

Alan Booth (Wymondham College, Norfolk)
Claire Buxton (City and Islington College, London)
Anne Constable (Beauchamp College, Leicester)
Corinne Harper (Burntwood School for Girls, London)
Dai James (Ashfield School, Nottinghamshire)
Mary Moran (Kingsway School, Cheadle)
Barbara Mottershead (Shevington High School, Wigan)
Robin Pellatt (Bishop David Brown School, Woking)
Rob Petrie (Exeter St Thomas High School, Devon)
Richard Pinnock (Vale of Catmose School, Rutland)
Brian Russell (Dixons CTC, Bradford)
Kalvin Turner (Bosworth Community College, Leicester)

Project Team

David Perry - Project Director
Louise T Davies - Deputy Project Director
Antony R Booth - Assistant Project Director
Jim Sage - Assistant Project Director
Maria Kyriacou - Project Assistant

Acknowledgements

Our special thanks to all of the Teacher Fellows and their schools and particularly their colleagues, partners, friends and children who have supported them, whilst they were writing to meet all the deadlines.

The Royal College of Art Schools Technology Project wishes to extend its thanks to the following for their support and help in the writing of this book - Kathleen Lund (Chief Executive) and her colleagues at the CTC Trust, the Department for Education, Office for Standards in Education (OFSTED), the Royal College of Art, and their representatives on the Project Management Group; British Standards Institute, British Telecom, Ann McGarry (Centre for Alternative Technology), Ruth Conway, Mike Cusworth (Crosrol), Crafts Council, Steven Wetherby (De Roma Ice Cream), Bob Dore (Philip Morant Technology College), Duracell, Nick Lowe (Dixons Store Group), Terry Fiehn, Gill Greaney (Beauchamp College), Maggie Grey & Clive Grey, Ian Alred & Sheila Tomlinson (Halfords), Dr Steve Scott (HJ Heinz), John Hinchcliffe & Wendy Barber, Jim Houghton, Mark Hudson (Bishop Fox's Community School), Michael Diesendorff (Husqvarna), Chris Miller (Lyon Equipment), Andy Leigh & Mike Maynard (Maynard Leigh Associates), Marks & Spencers PLC, Mike Martin, Phil Martin (Dixons CTC), John Myerson, (London Borough of Richmond), Julie Pennell (Allied Signal), Steve Crawford (Peninne Foods) Andrew Waddington (Vero Electronics Ltd).

Contents

Acknowledgements	3
Introduction	5
THE CHALLENGES	**6**
Now you see it! - OPTICAL ILLUSION	6
Mini monsters - WORKING METAL	12
Folding and joining - SHEET MATERIAL	18
Eat Italian - PASTA PRODUCTION	24
Making music - MUSICAL INSTRUMENTS	32
Shining bright - TORCH MODELLING	38
Designs on ceramics - FISH DISH	44
Catching the wind - WIND GENERATOR	50
Flashing lights - ELECTRONIC TEXTILES	56
Windows on the world - MOVING DISPLAY	64
Dipping, dyeing, dribbling - COLOURING TEXTILES	70
Layer by layer - FRUIT DESSERT	78
Remember us - PROMOTIONAL PRODUCTS	84
DESIGNING AND MAKING	**89**
INDEX	**127**

Welcome to D&T Challenges!

This book has two sections. At the front of the book you will find 13 activities, each of which starts with a design Challenge. Towards the back of the book you will find a section on Designing and Making. This section will help you with all of the Challenges. Turn to this section when you need help with a problem:

- How can I find out what a person needs?
- How do I write a specification?
- How do I choose the right materials to use?
- What can I do when my team does not work well together?
- How can I present my work?

This section will give you help and information. Try some of the activities in this section to develop your designing skills. You can also try these at home or if you have some spare time at the end of a project.

Your Challenges

You will probably not have time to do all 13 Challenges, but your teacher will pick a course made up of those that are best for your class. All the Challenges ask you to design and make things. You will use all sorts of materials and components, including food, textiles, plastics, wood, metal and electronics. Sometimes you will work with two materials together – have a look at *Eat Italian* and *Remember us*.

Over this year you will improve the quality of the products you design and make. Your work will get better and more complex. You will be designing for people and situations which are less familiar to you. You'll have to research and test your design carefully to develop better products. Throughout this book you will find examples of designing and making from the adult world. Try to improve your work until it becomes more like these examples.

You will also be working in a design team. Practise being a good team member – this will help you now and later in life.

The thirteen activities

You will see the following headings in each project. What do they mean?

Your Challenge   Each activity is introduced with a Challenge. Your teacher will introduce the project and plan it with you. Try to look ahead as this will help you to plan your own projects by yourself later. There are important pictures on the page which will help your designing and making.

Why this activity is useful   Each project has been included in this book for a good reason. This box will tell you what important skills and ideas you are going to learn.

The broader picture . . .   Each Challenge provokes important questions. When we are designing and making products this affects other people, the environment and so on. Your teacher may discuss these questions with your class, or you may like to answer them for homework, or find out more about something that is important to you. How do these issues affect your designing?

To be successful   Check this box often as it will help you to do well in each project. You can also use this to help you to evaluate how well you have done at the end of the project.

Planning things through   You can't depend on your teacher to do everything for you. You should take some responsibility for planning your own work. These are important hints and tips for you.

Now you see it!

Your challenge!

Optical illusions are fun and very mystifying. Sometimes they are just **two-dimensional (2D)**, sometimes the illusion is of depth, making them **three-dimensional (3D)**. They have been used throughout history to decorate and inspire in paintings, buildings, clothing, etc.

You are to explore some common types of optical illusion and **your challenge** is to create one of your own and to use it purposefully in a design.

Why this activity is useful

- ◆ You will develop your graphical skills and techniques.
- ◆ You will practise using and developing source material in a design.
- ◆ You will find out more about the work of famous artists, designers and architects. Their work may provide you with inspiration for future Challenges.
- ◆ You will find out about a natural phenomenon and how to use it to create exciting images.

The broader picture . . .

- A common expression is 'What you see is what you get'. Is this true?
- Who sets out to make sure you do not get what you see?
- Optical illusions are designed to deceive. Is this dishonest?
- 'Magic eye' pictures are very popular in the 1990s and Escher posters were popular in the 1970s. Are optical illusions a passing fad?
- Find out what a mirage is.

To be successful

- ★ You will explore and experiment with known optical illusions. These will help you to understand optical illusions and find out about their origins and creators.
- ★ You will produce your own optical illusion, work to test and improve it so it works really well.
- ★ Your illusion will be original.
- ★ Your product will be well made and effective.
- ★ You might use the illusion you create in a useful product.

Planning things through

There are lots of short activities here which could be fun. Your teacher may divide the activities amongst your class but you could try others at home. Your teacher will tell you what materials and equipment are available.

You will be responsible for using your own time to do some research about optical illusions, and for developing your ideas to a good standard.

Don't forget that computer graphics might allow you to experiment with and change your ideas more quickly. When could you do this?

Make your own 3D glasses

One common form of optical illusion is to make something flat, e.g. a picture, look three dimensional (3D). A way of doing this is to create two different coloured images and then use some special glasses.

You can use the example on this page to start you off or you can make your own original design. You can make improvements to your design so that the glasses fit better. You can decorate your glasses to make them look more interesting.

Find out more about 3D glasses:

■ Where can you use them?

■ What are they useful for?

3D pictures

Use your glasses to add the third dimension, depth, to the pictures. Wear your 3D glasses with the green side over your left eye. Find out what happens. How can you see 3D images? Create a picture of your own or find some more pictures which you can use your glasses with. Your class or group could make a 'magazine' collection.

Make your own 3D glasses

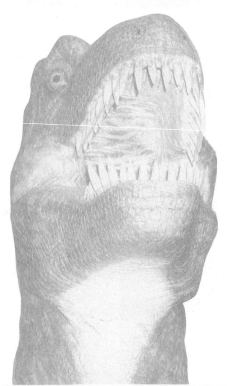
Use your 3D glasses on this picture

Double pictures

Look carefully at these pictures. See if you can find:

- the sad or angry man
- the old lady
- the vase or faces.

Look at the example of **metamorphosis** (changing shapes). The word is used to describe what a caterpillar does when it changes into a butterfly. Design your own visual metamorphosis. Will it have a special message? Where could you put it so that other people could enjoy it? Could you design something that would be suitable for the wall of a nursery school or hospital waiting room?

Impossible shapes

Look at these examples of impossible shapes. Escher was a very famous artist who was interested in people's **perception** – how we see things.

He developed many famous pictures through exploring this interest. Practise drawing some of these shapes. Find out more about Escher.

Your class could make a small display about all kinds of optical illusion graphics.

Ascending and descending

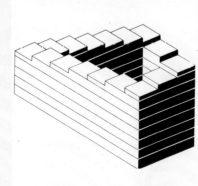
An exploded diagram showing how to draw the never-ending stairs

Lines and circles

Here are some more examples for you to try out on your friends.

■ Why do these images deceive your eyes?

■ Could you use these ideas in your design work?

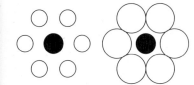

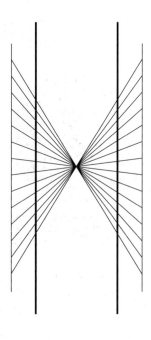

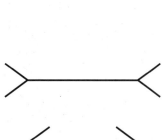

Arrest 1, Bridget Riley (born 1931)

Op Art

Op artists in the 1950s and 1960s used optical patterns to create illusions and effects. Look at this example. No matter how hard you look you cannot keep it still. Why does this happen?

Find out about the Op Art movement. Why was it called Op Art?

An example of Op Art

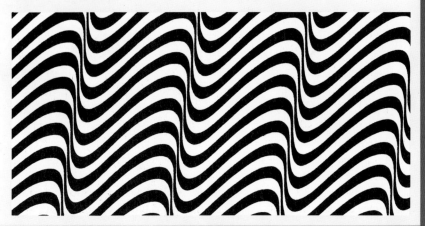

Tessellations

Look at these examples. What is a tessellation?

Create your own examples of tessellations. You will find this is easy to do on a computer using a drawing program, with COPY and PASTE functions.

Find examples of tessellations in buildings, or fabrics.

Once you have explored optical illusions and practised some of these techniques you may be able to incorporate an optical illusion into a product that you have designed.

What kind of product might be suitable?

You may want to capture people's attention with some unusual packaging. You can put an optical illusion graphic on the outer packaging instead of a picture of the product. This would be good where the product itself does not look very exciting, for example a music CD.

Record company logo

An optical illusion may be entertaining for people or, as decoration, it can improve an area such as a subway, or waiting room. This could be as a wall hanging, a painting, clothing, or ceramic tiles.

Mini monsters

Your challenge!

The world of insects is a fascinating one. If insects were any bigger we would surely think of them as monsters. The idea of monster insects has been used in many films. Artists often use figures from the natural world for sculptures which appear in public places or at the entrances of large buildings.

Your challenge is to design and make a mini monster big enough to be used as a door stop, paper weight or mini sculpture. The materials you use will be limited to steel tube and steel bar. Use the tube for the main body, if possible, and round bars to form features like the legs, tails, claws, antennae, eye stalks, etc.

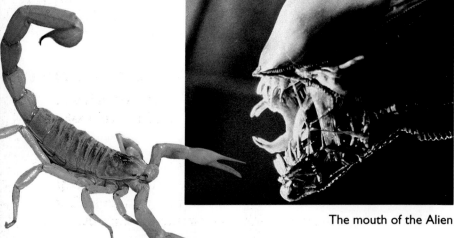

The mouth of the Alien

Why this activity is useful

This activity will give you some experience of using and designing for **mild steel**. Many of the products we find around us are made from steel, for example cars, bikes, gates, furniture frames and so on.

This activity will give you experience in:

- cutting, shaping and joining mild steel
- understanding the effect of heat on metal
- being adventurous and creative with your design
- making something you can keep, use and enjoy.

The broader picture...

- Some people design and make things because they enjoy doing so and not because there is a need for the product. (Although people may want to buy it – why?)
- What sorts of things are produced because designers enjoy being creative or because people want something to look at?
- Should we use precious resources for items that are not absolutely necessary?
- In some cultures novelty and decorative products are produced from reclaimed materials. Which materials can be used in this way?

To be successful

The way you form (bend and shape) your materials will make your design come to life – your product will look like a living creature, a real mini monster.

You will be proud of what you have made because it will be well crafted. All the parts will be smooth, with no sharp edges and no gaps where they shouldn't be. The surface finish will be attractive with no unwanted blemishes.

Planning things through

This is a short Challenge. Be sure you know exactly how much time you have. Your teacher will take some of the time to show you new processes (e.g. brazing).

You will be responsible for:

- thinking carefully about the order in which you do things because you will need to share equipment
- planning your work so that you can get on with something else while you wait to use limited equipment
- allowing enough time for finishing your work properly
- keeping all sketches and notes in your folio.

Understanding the material

See the 'What is steel?' worksheet to find out where it comes from. Before you start designing, you need to experiment with steel to find out what you can do with it. Your teacher will be able to help you and will have resource sheets with investigations for you to carry out.

Designing and Making: Thinking about materials 104

Two things you will have to consider are **compression** and **tension**. Try this experiment:
Take a piece of thick card and fold it as tightly as possible in one direction.

Now look at the inside of the fold – what has happened? Is it the same on the outside of the fold? Now try cutting part way through the same type of card (scoring). Fold it and compare it with the first piece you folded.

folding without scoring
signs of stretching under tension

a scored fold
scoring relieves the tension so the fold is sharper
You will see the difference very clearly if you try this with thick plasticine.

Two test folds in thick card

■ If you cut part of the way through a steel tube, is it easier to bend?

■ What are the disadvantages of doing this?

■ What happens if you bend a thin piece of mild steel too far? And a thick piece?

■ How can you prevent this happening?

Designing my monster

■ Look at pictures and photographs of real insects to give you some ideas for your design. Magnify them or look at them under a low-power microscope.

■ Which monsters do you like? What are the major parts of their bodies? How could you make them?

■ You may want to make a collection of pictures to put in your design folio.

■ What is your mini monster going to be used for? Make a decision about this because it will affect your design. What other things must you consider?

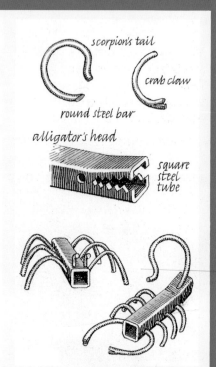
scorpion's tail
crab claw
round steel bar
alligator's head
square steel tube

Designing my monster (continued)

- A paper weight has to be a certain size and must not damage the paper. Why?
- An ornament needs to look good, even a monster! Why?
- A door stop needs to be a certain weight or to grip the floor surface. How?

Once you have chosen your monster and the product you are going to make, you will find it useful to sketch your idea and to begin to plan how you are going to make it. Do this by adding annotations to your sketches.

Finding out about ornamental steelwork

When you see ornamental steelwork, such as gates, railings or garden furniture, look at it carefully to see how the pieces are joined together. Do you know how ornamental steelwork is made? Is it all made individually by hand or can it be mass produced?

Is there a blacksmith near you? Where do the ideas for the blacksmith's designs come from?

Case study – The Queen Mother's Gates

The Queen Mother was presented with these gates in honour of her 90th birthday. It is a living sculpture in metal with a central screen depicting a tree of life with the Royal supporters, the Lion and Unicorn, together with various birds and animals representing her Majesty's lifetime and interests. It also includes symbols of our nation and the Commonwealth. Giusseppe Lund and David Wynne designed and built the gates which form a grand entrance to one of the most famous parks in the world, Hyde Park in London.

Making my monster

You will learn how to cut, shape and join mild steel. To do this you will be using special tools.

You may have already worked in mild steel. Your teacher will be able to show you this if it is new to you and will have resource sheets that give you more information.

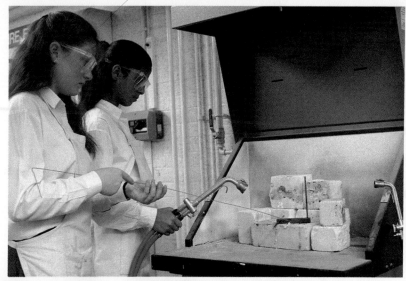

Heating the steel before joining the parts

Cutting the steel

Using a tube as a jig

Shaping the parts of the monster

Finishing my project

What does your monster look like when you have made it? How can you improve its appearance? What will happen if you leave it as it is?

Look at a range of steel products to see what has been done to the surface. Steel can be coated in various ways to protect it, make it look more interesting or preserve it. Some jewellers and sculptors deliberately **oxidise** (rust) their steel to give it an interesting finish. It can then be sealed with a clear lacquer spray.

Various surface finishes on steel products

Rusting used as a decorative finish

Looking back

When you have made your mini monster, look at it carefully and ask yourself whether it looks like the real thing. Is it a lively and exciting design? Are you pleased with the outcome? Is it well enough made to be satisfying to look at for a long time in the future? Is the product finished well? How does your product compare with the ornamental steelwork shown on page 15?

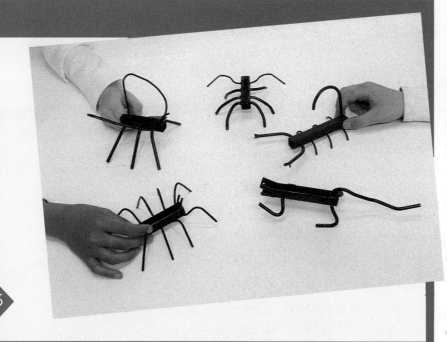

Designing and Making: Evaluating 115

Folding and joining

Your challenge!

Many products are made from a single sheet of material. Flat sheets can be folded to form a 3D shape to make furniture, machinery, packaging etc. Cardboard, metal, plastics or other materials can be used but some, though they bend, cannot be folded tightly.

Objects made of folded sheets can be very light and rigid, despite using little material. Folding flat sheets to form 3D objects can be more economical in production than joining separate sections together.

Complicated shapes can easily be modelled in card or on computer screens.

Your challenge is to design and make a product mainly from a single sheet of material aimed at the young teenage market and capable of being easily batch produced.

Why this activity is useful

- This Challenge will develop your modelling skills in creating 3D forms from 2D materials.
- You will learn how sheet material can be made into rigid, strong forms.
- You will consider how designs can be developed so that products are easier to manufacture.
- You will learn how to find out about and design for the needs of a specific age group.

The broader picture...

Some products are now aimed specifically at teenagers. Has this always been the case?

When designing products we sometimes divide people into 'types' but do you think this is accurate or fair? (See the case study on page 20) Can you identify how many different 'types' there are in your class? How do you decide?

Market researchers put people into categories for advertising new products. Do you think the categories they create are useful starting points for designers?

To be successful

- ★ You will need to identify a potential teenager group/type to design for, identify their needs and write a clear design specification for your product.
- ★ You will produce a variety of ideas and model your best ideas, possibly in card.
- ★ You will show how you developed your first ideas and improved them.
- ★ You will manufacture a **prototype** from the sheet material and evaluate it against the design specification.

Planning things through

Check that you know what machinery and equipment is available for your use and how you might need to modify your designs to make them easier to manufacture. When you use machinery you will be responsible for the safety of yourself and others in your class.

When you have produced your model, draw a diagram which shows the order in which you are going to manufacture your product, taking safety into account.

Discuss your model and plans with your teacher before you begin manufacturing.

Finding out more about teenagers

Designing and Making: Finding more information 91

Understanding who you are designing for is very important. There is no point in designing a product that no one will buy. You need to find out about the users' needs, interests and preferences.

Carry out a simple survey to gain more information. You need to design your survey carefully to help you choose the right people and to ask the right questions.

This simple client profile collage might help you to decide whether your design proposals are suitable. Would a 13 year old boy who was interested only in sport and computers be likely to spend his pocket money on a desk tidy? He might if it was produced with a sporty look.

Case study – Dixons' client profiling

How Dixons categorise their customers into 'types'

Dixons, the High Street electrical store, produced these cartoon images of a number of **stereotypical** shoppers based on their market research. This helped them to choose the products they sold and the positions they placed the products in the stores. Which cartoon fits your teacher or other adults you know?

IMPULSIVE SHOPPERS

Personal characteristics
★ Young (less than 35 years old)
★ Female
★ Single/married with young kids
★ High disposable income

Behaviour
★ Enjoys buying on spur of moment
★ First to try new products
★ Prefers well known brands
★ Doesn't bother to look for lowest price

Buying priorities
★ Product
★ Service (stock availability)
★ Price

INDEPENDENTS

Personal characteristics
★ Middle aged 35-44 year olds
★ Equal male/female
★ Single or married without kids
★ Upper/middle white collar

Behaviour
★ Technology aware
★ Not first with new products
★ Prefers well known brands
★ Wants value for money
★ Don't believe retailers give impartial advice

Buying priorities
★ Product
★ Price
★ Service

ASPIRING

Personal characteristics
★ Very young (18-24)
★ Male ★ Single
★ Working class

Behaviour
★ First to try new products/technology
★ Lowest price is the most important
★ Likes store credit
★ Likes to shop

Buying priorities
★ Price
★ Product
★ Service

Clarifying the task

Designing and Making: Identifying needs → 90

You will need to think carefully about the sort of products a teenager might wish to have which could be made from folded sheet material. Look at the picture of this teenager's bedroom. Can you see possibilities for designing new products?

One way to get ideas might be to look very carefully around a real teenager's bedroom for ideas. It may be better to look at someone else's room because you sometimes do not notice the opportunities for change in a place where you spend a lot of time. Sometimes you just get used to the room as it is.

Here is a list that Malika produced after she visited a friend's bedroom. Malika only looked at the obvious problems. If she had looked inside the wardrobe or talked to her friend's family she might have found many more.

Testing ideas on others (Group crit.)

When everyone in the group has looked at the sort of products a teenager might wish to have, share your findings. Sharing ideas at this stage is often very valuable. Decide what sort of products might be suitable for manufacture using sheet materials. What properties would be necessary to make each product successful? (e.g. Strong enough to hold a ... Flexible enough to clip on ...).

Developing a specification

Designing and Making: Specifications → 92

As well as designing for a customer's likes and dislikes you will have to think of the other design features it should have:
How will it work? Does it have to be strong? Is colour important? You must also think about designing for production. Is it going to be made by hand, 'one-off', or produced in large numbers?

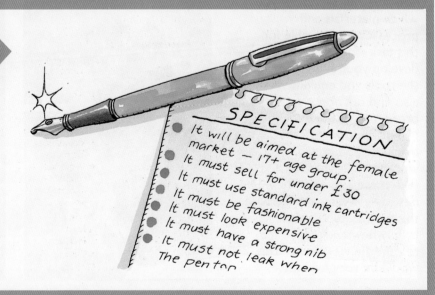

Thumb-nail 3D modelling

Design folios often contain thumbnail sketches (quick drawings which are not much bigger than your thumbnail). You can produce thumbnail sketch models to include in your design folio.

Often paper or card modelling is used to test an idea. Instead of sketching you can use scissors or a craft knife to cut out and produce some very different ideas.

You will need a selection of stiff coloured paper, scissors and a glue stick. 'Popset' paper is ideal as the colour is in the paper, unlike Poster paper where it is on one side only. Work quickly and try to make lots of models in a short space of time.

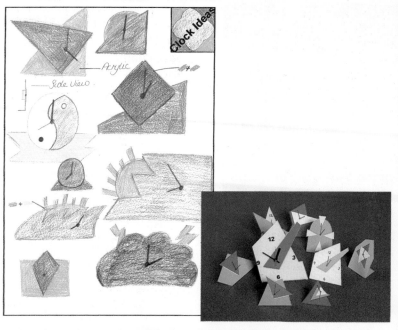

Both thumbnail sketches and 3D sketch models can be seen in this work from students. See how much easier it is to understand the models

Understanding the materials

Your teacher will show you which materials and processes are available for this task. You will need to develop your skills in using the tools and equipment and also find out about the properties of the materials. Manufacturers have specialist machines and skilled workers. Designers have to take these into account when designing products otherwise the manufacturer would have to buy new equipment and retrain the workforce. This would be very expensive.

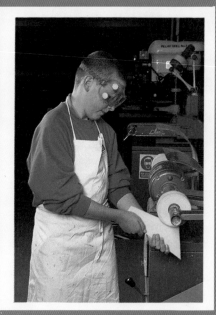

Case study – planning for manufacture at Crosrol

Here at Crosrol the sheet steel has been cut using a computer-controlled punch press. This machine is quicker, avoids waste and means fewer components need to be stored. A computer controls a small square punch which moves over the surface of the steel, cutting out the individual flat shapes. A large number of shapes can be cut out. They are drawn using a CAD package elsewhere in the factory.

Products are folded into their finished form using a computer controlled press brake. The computer monitor shows which stage of folding is next and the computer makes sure the fold is in exactly the right place. A few years ago, both of these operations would have been done manually.

Crosrol manufactures large machines for carding cotton. They can make a number of different models but only make machines to order. The computer shows how many parts are required for the particular model being produced. The correct quantity of parts is made and each part is finished just in time to be fitted to the machine. This sort of planning is often called 'Just in Time' and it saves a lot of storage space. Can you think how this could save money for the company?

Producing a simple plan

Designing and Making: Planning manufacture 118

You should produce a simple plan to show the order of the stages of manufacture. Add notes to show that you have considered the safest methods of working and the precautions you would adopt. You could do this for homework.

CAD/CAM

If you wanted to make a quantity of these products you might consider using computer-aided design and computer-aided manufacture (CAD/CAM). The changes you would make to the design to ease production would depend on the software you use, the size of the machining area and the cutting tools used.

A milling machine controlled by software, such as Denford's Mill Cam Designer, would be suitable for both cutting out the overall shape and engraving the details. However, there are other pieces of CAD/CAM equipment which might also be useful.

Eat Italian

Your challenge!

Italian food is very popular and ready-to-eat meals can be seen in every supermarket. There is a wide range of dishes using typical Italian ingredients – meat, mozzarella cheese, tomatoes, herbs. Pasta is especially associated with Italian food and there are many different shapes to choose and use.

A manufacturer wants a new fresh pasta dish for its cook-chill line. **Your challenge** is to design and make a prototype of a typical Italian dish.

You might be able to simulate a production line to see how your dish would be made in larger quantities. You may also design or use some special equipment to help in the production of your dish.

Why this activity is useful

- You will find out about the types of ingredients which go into Italian dishes, and how pasta is made.
- You will experiment with different ingredients, combining them to make a new product or improve an existing one. This is important for future new product development.
- You will think about how production systems are designed to produce a dish on a larger scale.
- You may work with other materials as well as food.

The broader picture ...

Many people think Italian food is just pasta and pizzas. Why do you think these are the most well known? How could you change people's ideas about this?

Why do people like to have so many shapes of pasta? Pasta can help towards healthy eating goals. Can you explain why?

Different flavours of pasta (e.g. lemon and pepper, sun-dried tomato) are being produced. What are the reasons for this?

To be successful

You will need to:

★ use your research into traditional Italian foods and existing products to design and make a new pasta dish

★ design and produce your own pasta shapes

★ explore plenty of ideas so you will finish with one that is excellent

★ think carefully how you can combine ingredients to make a dish which will appeal to consumers

★ understand how your prototype dish could be made on a larger scale.

Planning things through

Investigating Italian foods and recipes is important. You could work in groups and share your findings. Doing this for homework will save time.

Don't forget to leave time for trying different ideas, tasting and evaluating them before deciding on your final idea.

Your teacher will help organise the production of your dish and how you can 'scale-up' production to handle larger quantities. Be sure you know what time is allowed for this.

Finding out about Italian food

Before creating your new pasta dish, it is important to find out what Italian food is like, what ingredients are used and, in particular, what pasta is.

Italian cooking provides some of the most colourful and flavoursome foods in Europe. What makes it more interesting is that it varies from one part of Italy to another and each region has its own speciality dishes.

How can you find out more about Italian foods? How can you share the information with your class?

When food manufacturers such as Pennine Foods want to develop a new pasta dish, they send the chefs from their test kitchen to Italy. They go to different restaurants and try lots of dishes. When they return they try to reproduce the best of what they tasted and make it in large quantities for sale, not to be served fresh in a restaurant. But they want the dish that you buy in the supermarket to taste like the very best restaurant food.

Finding out about pasta

Pasta is a flour and water paste which comes in many shapes and sizes. It can be bought fresh or dried. Some pasta is bought in cans with the sauce already added. But all pasta is made from durum wheat semolina.

You could visit an information centre, the cookery section of a bookshop, a pasta shop, Italian Tourist information or factory as well

Fresh pasta is made with eggs. It is soft and should be eaten as soon as possible. Dried pasta, which is hard, is more popular because it can be stored for long periods. There are over 600 different varieties of shape and there can be flavourings added such as spinach or herbs.

Find out about pasta:

■ What is durum wheat semolina?

■ How is pasta made?

■ Name some varieties of pasta and collect some shapes.

■ Why are there so many different pasta shapes?

■ How should pasta be cooked? What happens when pasta is cooked?

■ What is added to make the colours different?

■ Describe the types of fresh pasta that are available.

Product evaluation

Look at the range of ready-made pasta dishes that can be found in your local supermarket. Record your findings as a chart like the one here.

You can also carry out a tasting session with different types of ready-made sauces. This will help you to find out what people like.

Designing and Making: Ideas from existing designs → 94

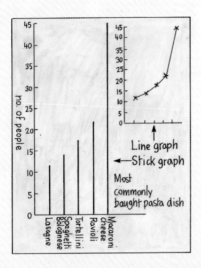

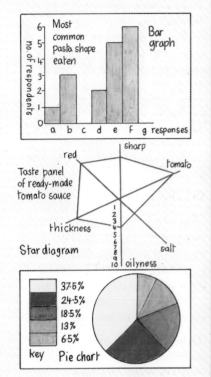

Taste panel for Heinz Pasta Soups

Graphs and star diagrams can be used to show your results

Developing your design

Designing and Making: Developing ideas → 95

To make sure your outcome is successful, it is important that you consider all these questions as you design your dish:

- What foods and flavours make a dish Italian-style?
- What qualities does a sauce for pasta need?
- How can you scale-up a recipe to make larger quantities?

The manufacturer needs to make sure that customers will buy the product. So, keep these questions in mind when you think of your first ideas, try them out and then evaluate them:

- Will the product be interesting, different?
- Who is it made for – people like your family or someone else?
- Will it be healthy?
- Will it look good? Will it look good and taste good when reheated?
- Will it be easy to use?
- How can you make sure each portion will taste exactly the same?

Case study – Heinz pasta soups

H J Heinz & Co Ltd – developing a range of pasta soups

Steve Scott (Development Group Manager) received information from Heinz's marketing department when they were developing a new range of soups.

Heinz soups have always been market leaders. Consumer research showed that there was increasing interest in soups containing pasta because people were eating more Italian foods. The development group were given guidelines to follow:
- Tastes and pasta shapes should be aimed at adults.
- The taste and recipe titles must be interesting but not too exotic or unusual.

Other requirements were identified.
- To develop and launch five pasta soup varieties:
  the lead variety to be an improved minestrone,
  the second variety to be a traditional chicken noodle,
  the other three varieties to be developed through the chefs' knowledge and creativity.

- Product characteristics:
  the texture should not be soft or slippery,
  a variety of pasta shapes and colours should be used,
  herbs should be used with the pasta,
  pasta can either be filled or unfilled.

The development team of food technologists and chefs experimented and Heinz now manufacture a successful range of pasta soups which can be bought in shops and supermarkets.

Manufacturing on a larger scale

Designing and Making: Making products in quantity 120

Describe the differences between these three manufacturing situations. How will the recipe and processes change in each one?

When a product is made on a large scale, the raw ingredients and the processes used are sometimes very different. If you are making a single dish you will use approximate measures such as one onion, one can of tomatoes, etc. and you will make it by hand, checking the taste as you go. A food manufacturer has to follow strict safety and hygiene procedures when they are making something which is going to be bought and eaten by the public. They also have to make sure that the quality is good in every single pack they produce. Each one has to taste and look the same.

It is a food technologist's job to make sure that the dishes produced on a large scale are as good quality as the prototype from their test kitchen. How do they do this?

Controlling the raw ingredients

A simple ingredient such as tomatoes can vary enormously.

Manufacturers choose their ingredients carefully and monitor them so that they are supplied with particular varieties to ensure good results.

PASTA PRODUCTION

Manufacturing on a larger scale (continued)

Controlling the process

At each stage of manufacturing a pasta dish, there is a complex plan to make sure each portion has the same ingredients and is made in the same way. This is called quality assurance. For example, there will be a measured amount of sauce and pasta placed in each dish and this is often checked by a computer.

Using machines

Manufacturers also use large machines to produce multiple copies of the same product. As long as the same ingredients and processes are used, each batch will always come out the same. Manufacturers use machinery to make the pasta dough and the pasta shapes. When we make things by hand they do not always come out the same every time!

A food production line at Pennine Foods

A large scale pasta machine with dies to form different shapes

Production line simulation

Have a go at setting up a production line and quality assurance system so that all the dishes come out the same. You can do this with your classmates.

Designing and Making: Working in teams 109

A commercial production line   Students working in a production line

Production line simulation (continued)

You may also want to try to design and make a piece of equipment or system which will help you do this.

Here is an example of a ravioli board from Pennine Foods' test kitchen. This would help you to get all your pasta shapes the same each time.

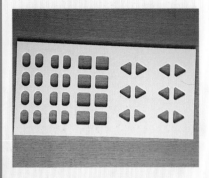

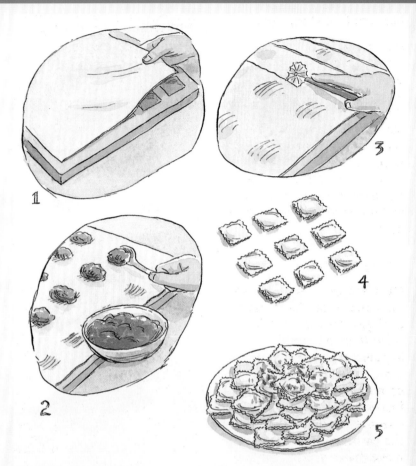

The board can be made from nylon or acrylic sheet. You can make your own personal ravioli shape. Design shapes that you would like for your ravioli. Cut out the shapes in a nylon or acrylic sheet. Your pasta can be rolled out and placed over the board. The pasta will sink into the cut out shape. A measured amount of filling can be placed in each one. Add a second layer of rolled pasta and your ravioli can be cut out.

 Sterilise the board before using

Here is an example of a control system. This will make sure that you put the right amount of pasta and sauce in each dish. It will warn you if it is underweight or overweight. This kind of system is used on a production line in a factory. Any product which fails the weight test is pushed off the production line by a pneumatic arm.

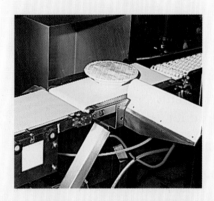

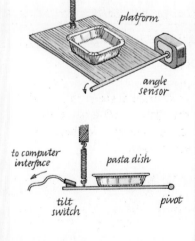

Making music

Your challenge!

For thousands of years people from different cultures all over the world have been creating music to express themselves. Instruments have developed considerably – from hitting a log with a stick, to complex mechanisms like the piano, and to programmable electronic instruments.

All instruments rely on the fact that sound is created by vibration. Whether an instrument needs to be hit, plucked or have air blown over it, vibration creates the sound.

Your challenge is to design and make a musical instrument that can sound at least five different notes. With your teacher you will agree what materials or components you are limited to and whether your instrument is to have a particular purpose.

Why this activity is useful

Most people enjoy trying to make a tune or rhythm on a musical instrument.

This activity will enable you to:

◆ create sounds and alter them to make different notes

◆ understand how the choice of materials and the accuracy with which an instrument is made will affect the quality of the finished product

◆ learn more about different cultures and times

◆ appreciate how important high quality of making is to an instrument.

The broader picture...

Why do you think music and musical instruments are so important in every culture? Even where music has become high-tech, using electronics to simulate sounds, great emphasis is still placed on the traditional aspects of making musical instruments. Is it important that these traditional methods continue?

What materials are used in different cultures to make their instruments? Why? Why are people prepared to pay such high prices for hand-made instruments?

To be successful

You will make an instrument which plays at least five different notes. To do this well you will:

★ experiment with making sounds in different ways and decide how you are going to create sound

★ make test pieces to know your final design will work well

★ choose your materials carefully

★ make your instrument accurately

★ tune your instrument so that your five different notes work together well.

Planning things through

You will need to spend some time making sure you understand how musical instruments work. Look in your library or resource centre. Will your music teacher or another musician come and talk to your class?

Make sure you know how much time you have to complete your activity. Make sure all the materials you need are available.

Take care when you are working out sizes and work accurately because this will save a lot of time when you come to tune your instrument.

How sounds are made

There are many ways of generating a sound. You can use different objects and materials to make a sound.

Do you get different sounds if you hit with different things?

What happens if you blow down or across a tube?

What happens if you put a blade of grass or thin strip of paper between your thumbs and blow through them?

How are these sounds made? What materials can be used?

Complete a chart like this one (right) for at least four different instruments. Share your information with your class. You should look at a wide range of instruments from different times and cultures.

How is sound generated electronically? Which instruments use similar systems to these?

Choosing an instrument to make

When you have experimented with making sounds and looked at instruments from different cultures and times, you will need to make some decisions. You will need to discuss these questions with your teacher:

- What type of instrument are you going to make?
- Why you have chosen this?
- What materials are you going to need?
- How will they be joined?

Communicating your design

> Designing and Making:
> Ideas from existing designs 94

You will need to draw your design for an instrument very accurately. You may design an instrument that is based on an existing one, in which case you may need to design only part of it.

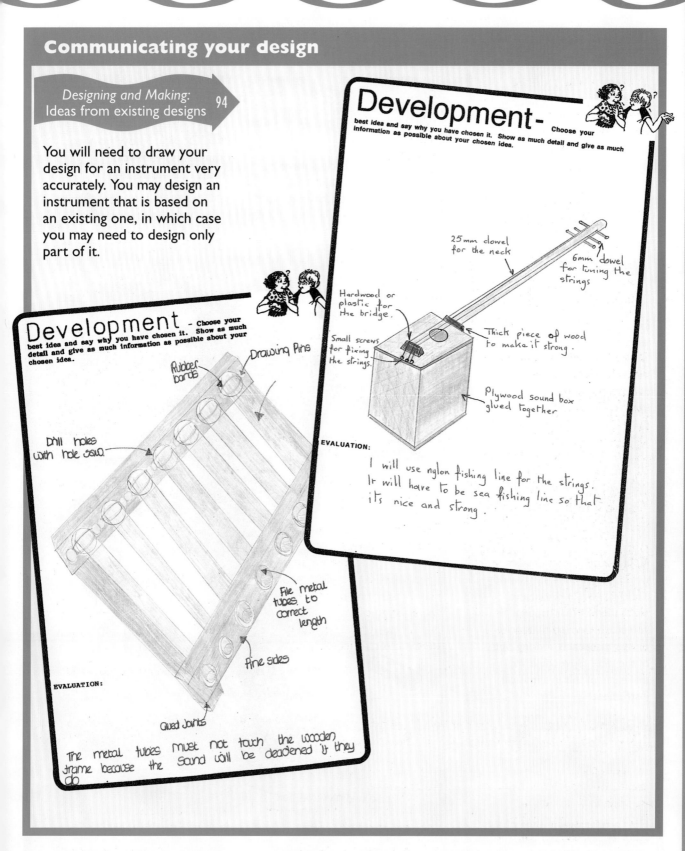

You may find it helpful to make a **mock-up** of your instrument. This will help you to test your design, see if your ideas will work and plan the making stage.

Mock-ups

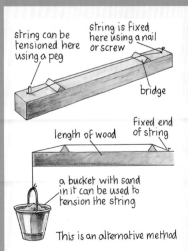

Experiment with strings. Use a length of wood with two 'bridges'. Try this with string, nylon fishing line, wire or an old guitar string. What happens when you alter the tension and pluck the string?

Hold different length pieces of wood, that are the same width and thickness, in a vice and pluck them. What happens if you move them so that more or less material is sticking out of the vice? How could you make use of this in a musical instrument?

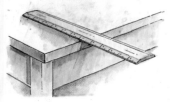

Hold a ruler on the edge of a table. Pluck the ruler as you move it in and out. What happens?

Planning to make your instrument

Designing and Making: Thinking about materials → 104

■ What materials have been used to make the instruments you have looked at?

■ What materials could you use to make your instrument?

■ Do you know how to shape and join your materials accurately?

■ How will you be able to tune your instrument?

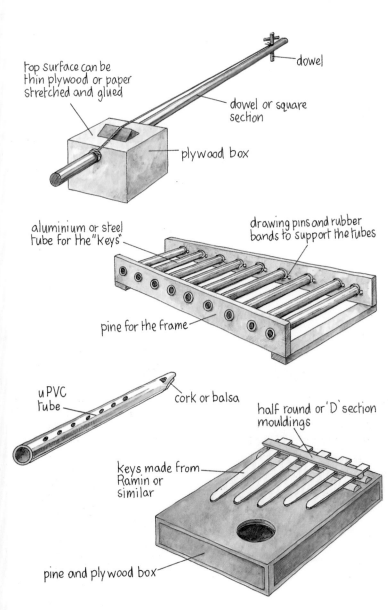

Tuning your instrument

You will need to find out how similar instruments are tuned.

Strings can be tensioned or tightened with a peg.

Wooden keys can have material removed from the underside.

Metal tubes and keys can have the ends filed.

The keys on a Sansa, or thumb piano, can be slid in or out to make them longer or shorter.

Tuning systems on different instruments

Evaluating

Designing and Making: Evaluating your ideas → 115

Part of your evaluation should be trying your instrument out. Did you aim to make your instrument play certain notes? Does it play them accurately? Your class could play a simple tune.

Shining bright

TORCH MODELLING

Your challenge!

Torches existed long before batteries. Batteries have made it possible to design a wide range of torches to suit every need and fashion.

Designers often try out new ideas by making **three dimensional models**. They can find out if their idea looks good, is the right size and shape and fits the purpose for which it is intended.

Your challenge is to design and make a **concept model** of a new torch or portable lamp to fill a gap in the market. It could be for use in leisure activities such as camping or cycling or perhaps an emergency light.

Why this activity is useful

- ◆ You will practise **investigating** what people's needs are and designing a suitable product for them to use.
- ◆ Your design will have to fit the people who use it so you will learn about **anthropometrics** and **ergonomics**.
- ◆ You will learn different ways to **model** your ideas.
- ◆ You will have the opportunity to design and make a new product.

The broader picture...

Disposable products, such as a single use camera are becoming more common. What are the benefits? What are the environmental effects of disposable products, especially torches?

SOS! – find out about Morse code and the use of torches and lamps for emergencies. What were torches before they had batteries? What has the word 'torch' got to do with spikes and twists? Does this give you any design ideas?

To be successful

★ You will need to identify a real purpose for your torch by analysing your market research results.

★ You will consider function and efficiency in your initial ideas.

★ You will develop your ideas, testing and modifying them using a variety of modelling techniques.

★ Your final concept model will be accurate and have a high quality finish. It will enable you to carry out consumer testing. Its ergonomics will be appropriate.

Planning things through

You can look at existing products and evaluate them for homework. You could also plan your market research at home. This will save you time. Allow time to discuss your ideas with others.

Your teacher will show you a variety of modelling techniques, but it is up to you to choose the most appropriate one for your torch design.

Agree with your teacher in advance what materials, tools and equipment are available and the time you will have for this Challenge.

TORCH MODELLING

Investigating products – what's in a torch?

In your class make a collection of torches, lamps and battery-powered lights.

Take one of these torches apart very carefully. Remove only the bits that are intended to come apart, for example, the batteries and bulb. As a class, record the following information as a chart, or as a series of labelled diagrams for the different torches.

- How does the shell stay together?
- How is the light reflected?
- How does the switch work?
- How many batteries have been used? What type are they?
- What materials have been used and why?
- How are the separate parts joined together and held in place? Are the joins hidden or made into a feature?
- Are there any special surface textures, e.g. for grip or decoration?

The right torch for the job

Look at the examples of different types of torches.

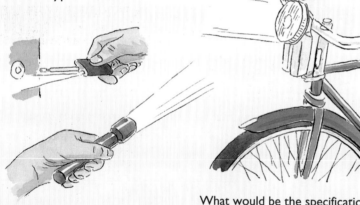

What would be the specification for each torch? Add to these examples

Pick one of these torches and carry out an **attribute analysis**.

Red book:
Identifying design attributes → 94

Imagine what the designers were instructed to do when they were designing that particular torch. Write out the specification that was given to them.

Designing and Making:
Specifications → 92

See how you could add ideas to develop these designs.

Who are you designing for?

You need to choose a person (or client) who needs a torch and find out what they would use it for. You must find out what their torch must be able to do to meet their needs.

This is the torch's primary (first and most important) function. You should also investigate possible secondary functions – what else might it do? Where will it be kept and does this create design needs?

Designing and Making:
Identifying needs → 90

Who are you designing for? (continued)

Carry out some market research which will help you know more. This research may involve observation of products in use and interviewing people.

You should use this research to decide:

■ Who are you going to design for?

■ What activity is your torch for?

■ What must the torch do to be successful?

■ How will you meet your client's needs?

This will help you to develop successful initial ideas.

Designing to suit your client

All products must be safe, comfortable and easy to use. You will need to think carefully how the torch will be used by the person you are designing for.

Designing and Making: Thinking about human factors → 103

Designing to suit your client (continued)

Ergonomics and anthropometrics

The size of the user is important to consider. Fitting a design to suit the user is called **ergonomics**. We are all different shapes and sizes, but for male and female, adult and child there is an average size for a particular age range and sex and you may design for the average. Alternatively, you may be designing for a particular person, so you might measure them.

These measurements are available for designers to use. The science concerned with measuring people is called anthropometrics (*anthropo* = people, *metric* = measure). Here is an example taken from the Compendium of British Standards for Design and Technology (PD7302). You should have this booklet in your school.

What data do you need to make sure that a torch is made to the right size to make it comfortable, easy and safe to use? How could you make sure your torch is the right size?

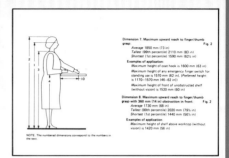

Designing and Making: Finding out more information **91**

Case study – the Petzl Zoom headtorch

Chris Miller of Lyon Equipment explains that the Petzl Zoom headtorch is the market leader in headtorches. It is used in many different activities. It is versatile and durable. Being head-mounted, the light beam always points in the direction the user is looking.

The Petzl Zoom is designed to be used in extreme conditions where failure could be dangerous, so its components and their materials are produced to very high standards.

The design of the Zoom headtorch features an adjustable ring which varies the beam from a wide flood to a narrow, intense beam. The bulbs are easily replaced. There is a halogen version of the bulb to give the torch a stronger light (up to 100 m).

The elastic straps are comfortable and easy to adjust so the torch fits the head or a crash helmet equally well.

Modelling your ideas

There are many materials which you can use for modelling, but your teacher will advise you on the ones available to you. You will be able to finish and colour your model so that it looks similar to the real product. If your model is accurate, you will be able to carry out consumer testing.

Designing and Making: Modelling your design ideas **108**

Quick modelling of different shapes can be done using modelling clay or Plasticene. This is a good method for developing a comfortable (ergonomic) shape for hand-held torches.

Modelling your ideas (continued)

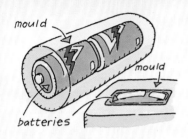

INSIDE-OUT: mould around batteries or shaped blocks to represent insides of the torch

OUTSIDE-IN: press into the modelling material to produce shapes designed to fit the hand

Modelling the torch body as a solid model using MDF or balsa wood:

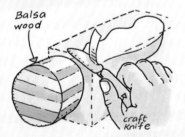

Modelling the torch body as a hollow model using Styrene or card sheet:

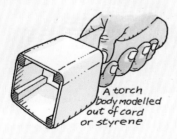

Complicated designs can be modelled using a number of different materials joined together to form a whole.

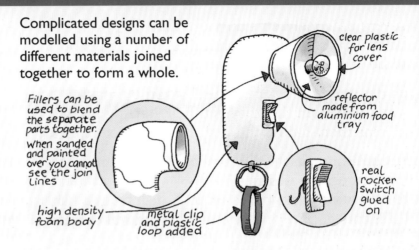

Here are a number of useful modelling items that you could find in school or at home:

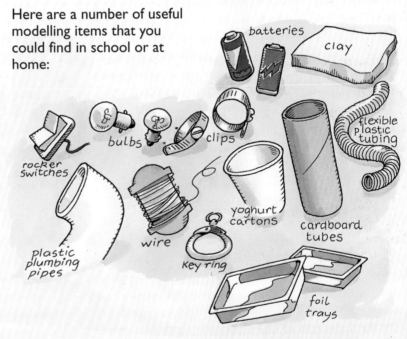

How could you use these in your modelling? What other items could you use?

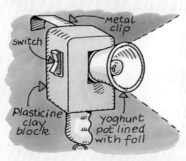

Try something totally different!

TORCH MODELLING

Designs on ceramics

Your challenge!

Even very practical items do more than carry out their function. For example, a dish has to hold things but it could be any shape. Where do we start?

Often designers study the natural world by making drawings and from this they get their starting point. This helps them focus, and develop their ideas quickly.

Your challenge starts with real objects and photographs to help you draw a design of a tropical fish and make a piece of ceramic ware based on it. A dish from a fish! You will concentrate on learning how natural objects can inspire designs and, starting from a 2D drawing, you will produce a 3D product.

Why this activity is useful

This activity will give you experience of how many designers work. It will help you to find visual inspiration in the future and then how to use these ideas. It will show you the relationship between drawing and making.

You will work in ceramics and understand how these materials behave. You can be creative and make something that you will enjoy using for many years.

The broader picture...

Why do some designers make objects which are purely decorative and serve no functional purpose? Why are some functional objects just used for decoration?

Why do people keep fish in both their homes and gardens?

Many underwater features such as coral reefs are damaged by leisure activities like snorkelling. How can people enjoy the beauty of tropical fish without damaging their habitat?

To be successful

★ Your drawing should be delightful in its own right.

★ You will show how you have used drawing as a starting point and developed it into an abstract design for a three dimensional product, without losing the quality of the drawing.

★ Your design should be original and creative.

★ Your product should be well made and visually attractive. You will be proud of what you have made and want to use it.

Planning things through

Find some pictures of tropical fish or use a local aquarium where you can sit quietly and draw them (it could be in someone's home). This is a simple homework task, but it will take some time. Don't be afraid of choosing unusual fish shapes.

Work at the drawing first by getting the correct shapes while you are observing the fish. Allow more time to add the finer details once the overall layout is right.

After the initial drawing you will need time to simplify and abstract the next drawing, starting to think in a three-dimensional way.

Collecting source material

You will need to collect some material on tropical fish. This is the starting point for you and many designers. What kind of material should you look for? Where would you find good pictures that you can work from?

Once you have collected a range of material, refine it and choose only the material that interests and stimulates you. The rest can be put into your recycling bin.

Try to look at live fish if you can

Choose pictures that are clear, detailed and colourful

How to draw fish

Your task is to use drawing as a way of understanding what fish actually look like. By observing them and drawing them you will see that they are not as simple as we sometimes think.

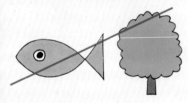

Look carefully at the photograph of your fish. Remember that the key to being able to draw well is to keep looking. Look at the object you are drawing, look at the marks you are making. Try and draw the basic shape of the fish as an outline, but in the same proportions as the original.

If you have difficulty in getting the proportions right, you could draw a rectangle around your fish and divide it into squares.

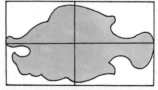

You can then measure your rectangle and scale it up to fit your paper. By dividing the rectangle you can look at each division and draw carefully the shapes you can see.

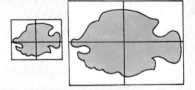

Once you are happy that the outline looks like your tropical fish, you can start to put in details such as fins, eyes, mouth and gills, making sure that you put them in the correct position and in the same proportions as the original.

Once you have this basic framework, start to work on your drawing. Look carefully at the relationship of shapes (pattern), texture and colour. Don't be afraid to start it again if you are not happy, it is always easier the second time around.

Design development

You will need to develop your work further. This may involve:

■ Simplifying your drawings in order to develop an abstract design which would be suitable for a piece of ceramics.

■ Repeating parts of your original drawing in order to create a design, a border, or a pattern.

■ Building up parts in order to create a relief.

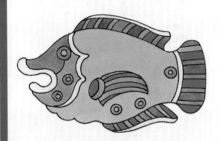
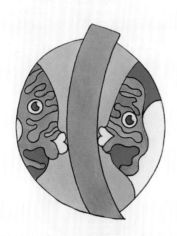

Once you have completed this you will need to think about how you are going to construct it.

Making templates

One method for constructing your work is hand-building using slabs of clay. In order to help you work accurately you can make some card templates. Templates are shapes or patterns which help you cut out accurately.

You can use tracing paper to copy onto card the shapes within your idea. Trace carefully from your drawing each individual shape, transfer them onto card using carbon paper and cut them out.

Making and constructing with slabs

There are many ways of building in clay. Many cultures use simple hand-built methods to construct individual items. These ceramics have an individual feel to them which is different for pottery that is cast in large numbers.

Mass-produced ceramics

One-off ceramics

Making clay slabs

A slab is made by rolling out the clay between blocks of wood (**laths**). Firstly the clay is flattened by hand on a canvas sheet. Then it is rolled out between the laths. The clay must be turned to encourage stretching and to prevent sticking. It is rolled from the middle outwards. The laths make sure that the rolling pin flattens the slab out evenly.

Roll from the middle outwards, using laths as guides to thickness.

Flatten by hand onto a canvas sheet.

Turn frequently to encourage stretching and avoid sticking.

Joining clay slabs

If you need to join pieces of clay, score on the surface and join with **slip** (watery clay).

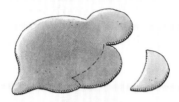

1. Cut out the slab shapes.

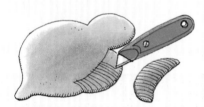

2. Score the surfaces carefully with a knife. Do not cut through the clay.

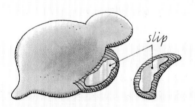

3. Spread slip on both scored slabs.

4. Sandwich the two slabs together. Wipe off any waste slip.

Evaluating your product

Designing and Making: Evaluating 115

How successful do you think you have been? (Look back to the *To be successful* box in the main Challenge.) Your own opinion is very important.

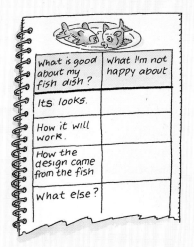

Work in small groups and discuss each other's work. Making positive comments and giving your reasons for any suggested changes will help you evaluate your product. What changes would you make if you had more time?

Case study

John Hinchcliffe and Wendy Barber

John Hinchcliffe and Wendy Barber were commissioned to design and make a panel by Bournemouth Museum to commemorate the opening of the new gallery. A decorative fish design on ceramic tiles seemed appropriate as the museum overlooks the sea.

They also design matching textile items such as tablecloths to complement their ceramic ware. You may also want to try this if you have time.

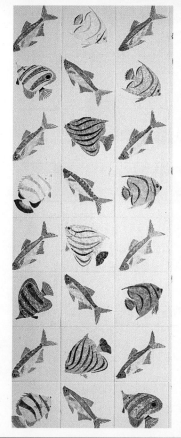

A final homework activity

Now that you have completed this project, draw a range of ceramics using fish as a starting point – you can be as outrageous as you like. However, you must complete this in twenty minutes. Make your drawings bold and clear, with smooth lines.

Catching the wind

Your challenge!

Can you imagine what life would be like if there was a permanent power cut? No television, no computer games!

In your home you use energy for many purposes, much of it in the form of electricity.

Most of our electricity is generated in power stations that use natural fuels such as gas or oil that are expected to run out one day. For this reason scientists and designers have been developing generators that use sources of energy which will never run out, such as wind and waves.

Your challenge is to work as a team to design and make a small device which uses wind to generate electricity.

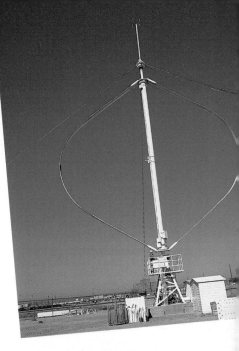

Why this activity is useful

- You will learn more about how electricity is generated and why it is necessary for us to find the most efficient ways of doing this. These are difficult questions which are being asked all over the world.

- You will learn how to use construction kits to model your ideas.

- You will be working as part of a team, so you will have to work well together to achieve results.

- You will also extend your experience of building structures.

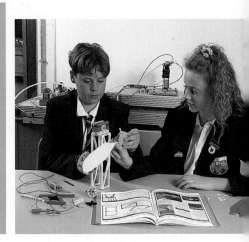

The broader picture ...

- What would you miss most at home without electricity?
- Look at where energy is used at home and how you might save energy.
- Windmills have been in use for a long time. Make a list of the purposes that wind power was used for. What else was used to drive mills other than wind?
- Why don't you see more wind generators in this country?
- Would you mind having a wind generator attached to your home? Would you like 20 of them behind your house? Or one on every roof?

To be successful

If you are to do well in this activity you will need to:

- make a wind generator and demonstrate that it works
- evaluate and modify your design to make it more efficient
- work well as a team member
- record your ideas and results carefully
- understand why it is important to save energy in the home and show how this can be achieved.

Planning things through

Beware! This activity is not just about making a wind generator – you will have to allow a lot of time for improving your design and finding a good use for the electricity. Work out an action plan with your team. Who will do what and when?

Check the availability of materials before you begin. Identify early on any materials that you have to find a source for.

Catching the wind

Windmills have been used over the years for many purposes. Look at the ones on pages 50 and 51.

What is each one used for? How do these windmills work? It will help you a lot later if you study old designs thoroughly and ask yourself what every part of them was for.

Are windmills useful today as a way of generating electricity? What are their disadvantages?

The windmill powers the light in this phone box

Designing a rotor – blades or sails

To catch the wind you will need to design and make a rotor. This is a shaft with blades or sails attached. To find the most efficient, you will need to experiment with different kinds, shapes and sizes.

Here are some examples that have been used successfully. How do they work? Can you think of any others? Why are mills powered by the tide more reliable than windmills?

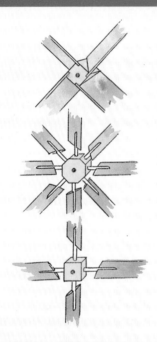

Fixing the blades to the shaft

In your teams think about the following:

■ What different rotor designs could you use?

■ What materials are available?

■ What could the shaft be made of?

■ Can you find anything to be a part-made sail?

■ What equipment is needed for shaping and cutting?

■ How can you test which is the best design? Could some team members make a simple test rig?

■ What is the maximum size you can make?

■ What is the best way to fix the blades onto the shaft?

Now add other questions you need to consider before you start work!

Raising the heights – designing a support system

Your team will need to design and build a tower to hold the shaft and sails high enough to prevent the sweep of the blade from touching the floor or table. Your tower structure will also need to hold all the mechanism that is to be attached to the rotor (look ahead to the 'Conversion to electricity' section).

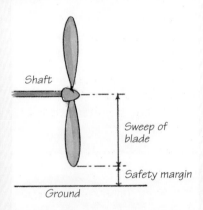

For a horizontal shaft, the sweep of the blade will dictate the height you need. You must measure this and allow a safety margin

Now consider which approach your team will use to gain sufficient height and make a strong enough structure. SIMPLE BUT STRONG – they are the watchwords.

Your team will need to think about:

- the materials that you will use
- the form and arrangement of the parts, including the shaft (horizontal or vertical?)
- how you will support the shaft so it can turn as freely as possible
- the weight and size of the blades
- the joining methods that you will use

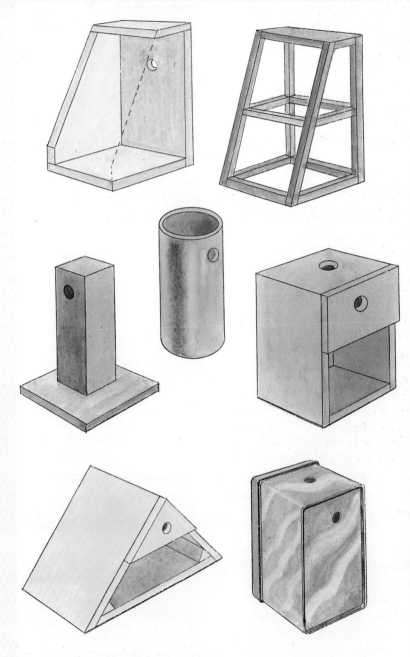

Tower-building methods

- the strength of the structure
- where any platform may be and if it will move.

WIND GENERATOR

Strong structures

A real wind generator has to be held in place high in the air and in strong winds. How are strong tower structures made?

All structures are made as light as possible to save money and not to waste materials.

There are three types of structure these are:

■ Shell structures
Car bodies, domes, snail shells and dishes are examples. Most large wind generators are mounted on shell structures. The strength of a shell structure comes from its shape. For the 'shell' to be tall and narrow will mean a tube or cone.

■ Frame structures
Electricity pylons and steel-framed buildings are examples. Most small wind generators and wind pumps are mounted on frames. The strength of a frame comes from being made up of triangles.

You could use construction kits or square section wood with gussets so that the many joints will not result in too weak a tower.

■ Slab structures
A third type of structure, e.g. a house, is made of slabs of material.

Red book: Identifying structures — 68

Conversion to electricity

The energy transferred from the wind into the movement of the rotor can be converted to electricity. This means that when the wind catches the sails it can rotate an electric motor which will then act as a generator.

The electrical output can be measured with the use of a volt meter or multimeter.

You will need to attach your rotor shaft to a motor. These are three possible ways

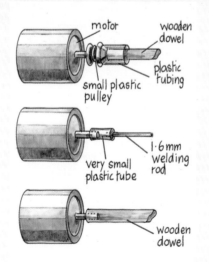

You will find that only a small amount of electrical power is generated, perhaps less than 10 milliwatts.

Speeding things up

The amount of electrical power generated depends on the speed of the motor/generator, which depends on the rate at which the sails turn. But too strong a wind could damage the windmill.

The speed of the motor can be increased by using gears between the sail and the motor/generator.

How do gears work?

The diagram shows two gear wheels, one with 10 teeth and one with 30. When the large wheel makes one rotation, how many does the small wheel make? As the two wheels take the same time for these turns the small one must be turning faster, which is what we want.

You may use construction kits, such as Lego Technic or Fischer Technic. Make a gear box to give a ratio of 1:3. You can then attach this box to your platform and sail. How much faster will your motor/generator turn now? This should give you sufficient speed to generate electricity to light at least one LED. With further gearing it should be possible to light a small bulb.

Modifying your design

You will need to carry out some tests to improve your first design. *Remember – useful tests require only one thing to be altered at a time.*

Designing and Making: How to test your products 116

Test different ways that you might:

■ keep the blades facing in the best direction

■ keep as steady a speed as possible

■ maintain a reasonable output

■ avoid driving the generator too fast.

Try these adjustments:

■ using blades of different designs

■ trying different positions for your blades

■ reducing friction

■ changing the gear ratios.

What else can you experiment with?

You can measure the output by using a voltmeter or multimeter and plot your results on a chart.

Remember to note every design change and plot the change in performance in your design folio.

The speed of the blades can be measured using a light sensor and sensor software, and plotted against power output.

Flashing lights

Your challenge!

Many more manufacturing companies adapt existing circuits and build them into new products than design completely new circuits. Imagine you are a designer working for this sort of company and have the responsibility for developing new products.

Your challenge is to build the circuit shown below to control alternate flashing **LEDs**, and then to design and make a new textile-based product which will make use of the circuit. Flashing LEDs make very useful **warning beacons** and this circuit can be fitted into a wide variety of products such as bags or safety clothing.

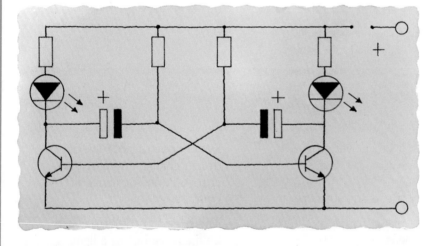

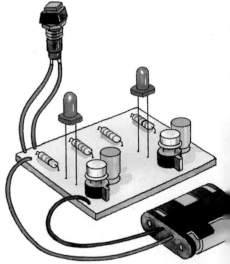

Why this activity is useful

There is a lot of interest now in finding new applications for electronics. This Challenge will help you develop skills and understanding in the field of electronics.

You have the opportunity to work on your own or as part of a production team and to learn more about industrial manufacture.

You will use your previous experience of designing and making to produce a product which will be as good as any you would find on sale in shops.

The broader picture . . .

◆ Can you think of different situations where flashing LEDs on clothing or other textile products might be used to improve safety?

◆ Some people think that battery powered products are not environmentally friendly. What do you think? What other sources of energy could be used?

◆ Many novelty products use flashing lights. Is it wrong to manufacture products just for fun?

To be successful

★ Quality is a very important feature of this Challenge. The product should be well made and your circuit should work.

★ You should be able to identify each of the components and explain what job they do in the circuit.

★ You should have a product which satisfies all the design criteria you have identified, including batteries which are easy to remove.

★ You should test the finished product and listen to other people's ideas on improving it.

★ Have fun and give pleasure, or be serious and give value to others through your product.

Planning things through

The activities in the next few pages help you to make a textiles product, but you can use other materials if you wish. Write a check list (a **specification**) for all the things your product needs to do.

When your classmates are producing the same circuit, you can organise a production line to make them more efficiently.

Think about how you will ensure quality control. You should produce a production flow chart for making the circuit when you have finished the activity.

How the circuit works

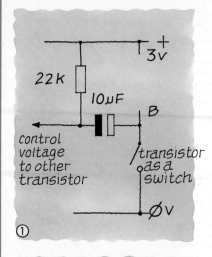

①

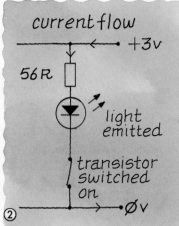

②

The circuit is in two halves, which have exactly the same components. The action of the circuit is easier to understand by looking at one half only. The transistor is represented by a switch. (Look at diagram 1.)

When the transistor is switched on it immediately makes the voltage on the 10 µF capacitor negative, which switches the other transistor off. Then the capacitor charges up through the 22 k resistor and the transistor until the voltage is sufficiently positive (+0.7 volts) to switch the other transistor on again. The other half of the circuit then repeats this sequence.

When each transistor is switched on, current can flow through the LED which makes it light up. The 56 ohm resistor is there because LEDs only need two volts to work. The resistor takes the remaining one volt of the supply voltage. (Look at diagram 2.)

The speed of switching is controlled by the values of the capacitor and resistor. If either is made larger the flashing rate will slow down.

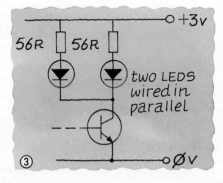

③

Building the circuit

You will need a printed circuit board (PCB). The components are soldered into the board starting with the resistors and ending with the transistors. The circuit board shown (in diagram 4) includes a diode so that the battery cannot be connected the wrong way around. The construction of the circuit could be organised as a production line. The LEDs, capacitors and transistors must be put in correctly. Refer to the fact sheet for details.

The LEDs could be connected to the PCB by lengths of insulated wire instead of being soldered directly to the board. It is possible to use more than two LEDs. Any extra LEDs should be wired in parallel with each other, and each should have its own 56 ohm resistor as in diagram 3.

This is a mask which will help you to make your PCB

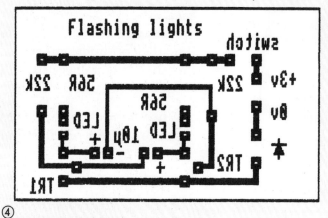

④

Designing an item to use your circuit – clarifying purposes

Your product will fall into one of two categories:

▪ something fun or decorative that is designed just to give pleasure

▪ something that performs a serious function.

Of course, if you are really imaginative, you might do both of these things.

A fun or decorative product

Explore ideas by brainstorming all that you can do with flashing LEDs.

▪ Where might they go in a garment or other textile article?

▪ Will they all be bunched together on a one-piece circuit board? If not, how will you make connections between the sections?

▪ Will there be a pattern to the flashes? Will it signal something?

▪ What colour LEDs will you use and why?

▪ Will there be any permanent part to the design (e.g. embroidered or dyed parts) together with the LEDs?

▪ Will the LEDs show all the time? If not, how will they be hidden?

Have fun, sketch freely. Only tidy up drawings afterwards. Add notes to them so the ideas can be understood (and so you'll remember them!).

A functional product

Find out what exists already that uses a circuit with similar functions. Remember, **flashing** is what is important. Flashes demand attention. This can help with safety; someone in danger can be noticed.

Search for situations where a product might help to keep people safe. Talk to a safety officer or other expert at a factory, shopping centre or any other busy place.

Ask: What do people do here that puts them in danger if they are not seen? Also, just look around and think about your own life – when are you at risk?

Of course, if a safety item is attractive to wear, people are less likely to leave it off.

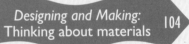

Designing and Making: Thinking about materials → 104

ELECTRONIC TEXTILES

ELECTRONIC TEXTILES

Case study – Halfords

Selling products designed to improve your safety

Cycle safety awareness has been rapidly expanding over recent years. Halfords is one of Britain's largest retailers of cycles and equipment and have seen a terrific expansion of sales for flashing cycle lights since they were first introduced in 1993–94.

Like many other responsible companies, they have to make sure that the items they sell are thoroughly tested before they go on sale. This is especially important with products which use new ideas or use existing technology in new ways. But sometimes the law lags behind.

The British Standards Institute (BSI) tests new products, particularly where safety is important. Products which meet their standards are allowed to show the BSI logo, the kite mark, on their packaging. Customers choose products with this recognised logo on it. How will you ensure that the products you make are safe and reliable?

In the case of bicycle lighting, for example, the Highway Code

HALFORDS

Halfords logo

gives very clear guidance. Did you know that flashing lights are illegal as the only light source when cycling after dusk? Shops, like Halfords, make this clear on the flashing cycle lights they sell as they have a responsibility to inform consumers about their products. They do this at the point-of-sale in a shop or on the packaging it comes in.

What research might you need to do to make sure that the products you make satisfy legal requirements?

Developing your design – using textiles

Identifying what the product will be used for and who you are designing for is an important first step. There are many small textile items which we wear but which you would not call clothing, e.g. bags, straps, bands, hats, bracelets, badges, scarves. The colour, shape and style will all be important, so will the type of fabric you choose for the product.

Think about:

■ how the product will look

■ how it will be used

■ how it will be made

■ what sort of details it will have.

Annotated drawings will help you to develop your ideas.

Red book: Drawing your ideas **98**

You can also model them in paper or a cheaper fabric to try things out or try

modelling a range of design ideas using a computer drawing program. You may also need to test and experiment with different fabrics to get the right one for your product.

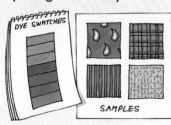

Patterns

You will need a pattern to work from. A paper or cheap fabric model would help you make this pattern.

When working with materials such as plastic or sheet metals you would use a template. Patterns must include information about seam allowances, darts and other features. A seam allowance, for instance, is the amount of fabric needed to support the seam: it will pull apart if the stitches are too close to the edge. This information is important because the fabric will usually have to be cut to a bigger size than the finished item.

You can use or adapt ready-made paper patterns. Or you may need to design your own pattern.

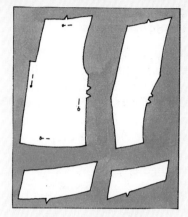

Here is a pattern for a child's waistcoat which has been designed to carry the flashing circuit. The designer has considered the position of the LEDs as well as a novel battery holder

Joining and finishing

There are many different types of seam used for different purposes. Your teacher may show you some which are suitable for your design.

Where there is an open edge on your design, such as the top of a bag, you will need to think about turning the edges and making a hem. There are many ways of hemming or binding such an edge.

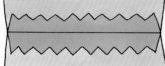

Types of seam

Joining and finishing (continued)

Pockets and reinforcements

If you are going to attach a pocket or a strap, you will need to practise this. Look at examples of different kinds of pockets. How are they stitched? How are they neatened? How are they strengthened?

Thinking about fastenings

As there are lots of fastenings available, some of the problems like battery access can easily be solved by making pockets with flaps.

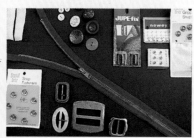

Which sorts of fastening could be used in your designs?

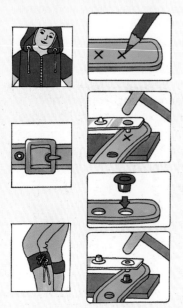

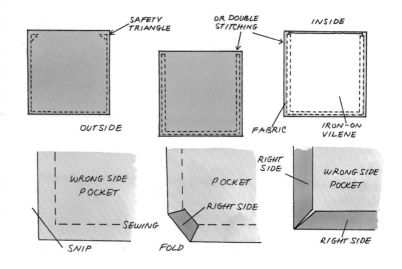

Buttons and buttonholes

1. Use buttonhole presser foot.
2. Mark buttonholes on fabric.
3. Place the fabric in the machine with the needle above the beginning of the hole.
4. Sew the first side.
5. Reverse.
6. Sew other side.
7. Machine will stitch the ends.
8. Cut the hole with the special knife.

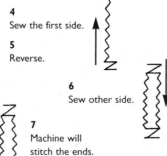
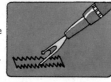

Buttons come in a variety of shapes. Sometimes this is just for style, sometimes this is for a purpose. What might the reasons be for choosing each of these designs?

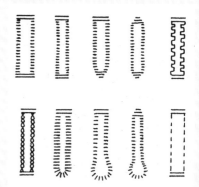

Husqvarna have many different buttonhole styles programmed into their machines. Your sewing machine may have other styles

Fine finishing

You need to incorporate your circuit into the product, (the LEDs, wires and batteries). You should consider how to do this carefully. You can make it an attractive feature of the product.

Decoration

You can choose to decorate fabrics in a number of attractive ways. Some of these you would need to plan before you make your product, others can be added afterwards. The decoration should suit the purpose and the user of the product.

You could use the computer to design the decoration and experiment with different ideas.

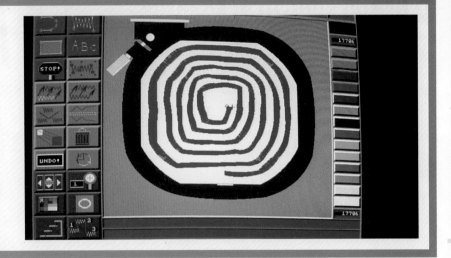

ELECTRONIC TEXTILES

Windows on the world

Your challenge!

On most high streets you will see a variety of food outlets, restaurants and take-aways, selling food from other countries. We enjoy the food of other countries, but how much do we know about their background culture?

Your challenge is to design an attractive, original, active shop-window display for a food outlet which promotes the culture of a country other than your own. Making it active (with moving parts) will help it catch the eye of people walking past.

You will work as a group, using a range of materials to create your display. You will need to research the culture of a country and select suitable features that promote it in an interesting and lively way.

Why this activity is useful

You will work in a group as a design team, and learn how to keep a log of your part in the team's work like professional designers.

You will work with a range of materials and bring these together in an active display. Making your display move will give you practice in working with electric motors and mechanisms.

You will learn about the richness of other cultures and communicate an 'image' of that culture. You will also learn about visual communication techniques.

The broader picture...

- Should companies make profits from other countries' food without knowing anything about the culture it comes from?
- Tourists are often accused of exploiting (taking advantage) of the people who live in the countries they visit. Is this fair?
- If you were to promote a cultural aspect of your region to a foreign visitor, what would you promote with pride? What local food might this include?
- What is the Advertising Standards Authority and is it necessary?

To be successful

You will

- work effectively and co-operatively in your team
- keep an attractive design folio, recording clearly your part in the team activity
- select appropriate images and suitable materials to promote the culture of your chosen country
- combine various materials into an eye-catching, active display which will be attractive to your chosen audience
- create an effective moving display that works well and looks attractive.

Planning things through

Your research into various cultures will need to result in a **mood board** conveying an image of the culture you have chosen.

You will explore mechanisms and control systems in class activities. Discuss, as a group and with your teacher, how you can employ systems and sensors in your design ideas. Allow homework time to identify sensors in various devices.

Experiment with alternative designs before you decide on your final proposal. Plan with your team how your group will share out the work. When will you meet as a team?

Creating a mood board

Look at the mood board illustrated opposite. It was designed to represent the 'spirit of the 1980s'. What do you think is represented by this collection of images and objects? What helped you to decide?

Create your own mood board to 'explain' your chosen country by collecting photographs, fabrics, papers, colours, small objects, examples of the written language, etc. and arranging them together to communicate to other people the culture of the country.

Designing and Making: Floating ideas → 90

Visual communication – what is it?

We communicate a lot through words but many ideas can be communicated more effectively visually.

The illustrations show two CD cover designs produced by the Pickwick Group. They are for compilation records including different kinds of music from the 1960s and 70s.

The designer, Phil Filipiak, selected these images from larger collages (such as the one at the top of the page) that he created to 'sum up' and communicate a feeling of each different decade.

Look at *The Two of Us* CD cover. Why do you think these images were selected for a love song compilation? What is the effect of the colour used? What is the mood of the 1960s that this makes you feel?

You now need to do this to create the mood of the country you are to represent.

Scale and colour are very important in your design work. Using space and colour effectively attracts attention.

Is your display being seen only from the front? If so, a layered design might be suitable. Sort your design into layers, three should be enough. Stage scenery designs often work like this, and so does the scene by Sam Smith below. Can you see the way the waves move?

Think about:

■ How are you going to create each layer?

■ What is the best way to create each layer?

■ What materials can you use?

■ Which parts are going to be moving?

Getting your display moving

What parts of your display will move? What mood are you trying to create? Are there any parts that would help to create this mood by moving, such as gentle wave movements, or dramatic pop-ups like a Jack-in-the-Box?

Can you include parts that would be moving in real life? For example, giant versions of Swiss Army knives are displayed in shop windows with the many blades moving in and out. These work like a series of layers which is a good approach for you to think about.

As well as *which* parts will move, you must think about *how* these parts will be made to move. These are best thought about at the same time, so you design something that it is possible for you to make.

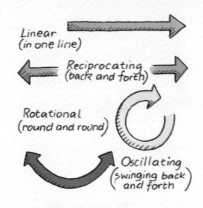

Mechanisms

Something made with moving parts is called a **mechanism**. Your mechanism might move in a number of directions, but beware of making it too complex.

There are four main types of movement. They can be used alone or in combination.

Find real examples of mechanisms for each of the four directions in the examples below

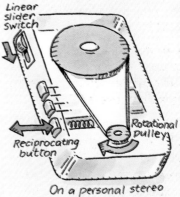

Designing your mechanism

Levers could be useful as they change the direction of movement, and can increase the force of the movement.

> Red book: Investigating levers — 63

You will also need to power your mechanism with an electric motor and you may want to use a computer to control it.

Your teacher may give you some sheets to help you design your mechanism.

Looking at real moving displays will give you ideas for your own

MOVING DISPLAY

67

Creating lettering

Lettering is an important part of your designing. Match the type of lettering (**font**) to the purpose of the lettering.

- ■ Must it be clear?
- ■ Will you design your own font?
- ■ Can you find a suitable commercial one?
- ■ Which words need to stand out the most?
- ■ Which the least?
- ■ Or is the impression it gives more important?

How are you going to set out your lettering?

Applying dry transfer lettering

Tracing from lettering sheets

Using the same font in different sizes

Changing letter styles from word to word

Lettering cut or ripped from coloured paper, newspapers, magazines

Letters cut out from magazines, balsa wood, polystyrene

Lettering in progress: on squared paper, drawn free-hand, drawn with 2 pencils at once

Use letters you have cut out or use the paper you have cut from

Overlap words

Use photographs behind text

Try computer generated designs

MOVING DISPLAY

Working in a design team

Designing and Making: Working in teams → 109

Working in teams can mean that several tasks are divided up and given to different members or all can do the same task. If you all work on researching as a team you will find more information.

Later each team member may need to take quite different jobs.

You will need to hold group meetings and make decisions as a team. You will need to plan individual tasks and to make sure that you meet deadlines. Keep records of what you have done and what you still need to do.

WORKS MANAGER

GRAPHIC DESIGNER

PRODUCTION MANAGER

TEXT DESIGNER

ELECTRONICS TECHNICIAN

TEAM PLAN

DAY 1 – Research: who is doing what?

DAY 5 – Report back research findings to the team.

DAY 7 –

MY TASKS

Research Mexican culture – feedback to the team in 4 days.

THURSDAY!

Dipping, dyeing, dribbling

Your challenge!

Colour is one of the most exciting parts of designing. We notice the colours of food, of clothes, of cars, of shampoo, of everything. We all have favourite colours and they can make us feel calm, happy, gloomy ...

Your challenge is to develop your own colourful design on fabric and use it to make:
- a bag, shorts or night shirt or
- a wall hanging.

Indian and Asian fabrics are highly decorative and there are different ways to dye and colour them. You will investigate some of these ways of colouring fabric:

- tie and dye
- batik
- block printing.

You could re-use fabric from old clothes, re-colour it and make it into something new.

Why this activity is useful

Colour has a major effect on our lives and this activity will heighten your sensitivity to it. You will learn about colouring fabric using traditional and modern methods.

There are many ways of decorating fabric and these **techniques** will show you how to decorate fabric in an interesting way.

The broader picture...

The colours we use often have important meanings. For example, people often wear black in Britain at funerals, and red usually signals danger. Find out about the meanings of as many colours as you can.

Emergency exit signs have changed from red to green. Why do you think this was done?

Christmas is usually associated with red and white. What other events have special colours associated with them?

To be successful

You'll practise colouring fabrics using a range of traditional methods. You'll use these skills and knowledge to design and make a colourful bag, shorts, night-shirt or wall hanging.

You'll choose a colouring technique to suit your design. The end result should be well made with accurate stitching and finishing.

Planning things through

- Decide what you are going to make before you begin colouring the fabric.
- Check which methods you'll have time to investigate.
- Which parts of this Challenge can you do in your own time?
- Will it help to work as part of a team to try more techniques?
- Find out the variety of fabrics and colours that you can use.
- Allow enough time to practise your dyeing and colouring techniques.

You'll need to check with your teacher what is going to be possible in the time you have for this Challenge.

COLOURING TEXTILES

Understanding more about colours

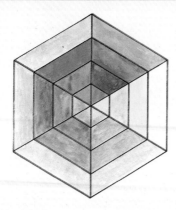

A colour wheel

Colours can create different atmospheres or feelings. Generally, the colours next to one another on the colour wheel **complement** each other, while the colours opposite **contrast** with each other or even clash.

Whatever you are doing you will see colours. Some colours are just so good they make your mouth water, some you really dislike – but we each have our own likes and dislikes.

When professional textile designers are working with colour they often get ideas from the world around them – from nature, from scenery, from buildings. Then they try out ideas on a colour board, mixing fabrics and yarns to match the colours they have seen.

Colour is one of the ways in which we express ourselves. Think how we can show support for a football team by choosing to wear their colours.

The colours people choose are said to reflect their personality. Can you tell from the colours people wear or have in their homes whether they are quiet, adventurous, or respectable?

Do the colours you choose reflect your personality?

Using natural dyes

Long before commercial powder dyes were used, people coloured fabrics using natural **extracts** from fruits, vegetables, flowers and plants. They used roots, bark, berries, grasses and leaves. Evidence of coloured fabrics has been found dating back to Ancient Egypt – 5000 years ago!

People cut plants up, then soaked them and cooked them in water. The coloured water was called a dye. The fabric was then soaked in the dye to change its colour. People used different plants to make dyes according to where they lived. In Scotland they used heathers and lichens to dye tweed cloth. In Mexico they use the dried bodies of the female cochineal beetle, which feeds off cactus plants, to give a bright red colour.

A woman crushing dead cochineal beetles to make dye

The most expensive dye ever came from shellfish. It took 12 000 shellfish to produce one gramme of purple dye – it was called 'Royal Purple' because only emperors could afford it.

This purple dye is occasionally still used by Mexican Indians. They pick the Purpura Patula off the wet rocks at low tide, squeeze and blow on the shellfish. This distresses the shellfish, which secretes a liquid onto a yarn held against it. Over the next three minutes, the liquid turns the yarn from transparent to dull yellow, vivid green and finally purple. The dyer returns the shellfish to the sea, only to 'milk' it again a month later.

In some traditional societies of the world people use **indigo** to dye things blue. They pay the utmost respect to the indigo dyepot, and are superstitious about it. Indigo is a difficult dye to use, so in Indonesia and Nigeria a chicken feather is hung over the dyepot to ward off evil spirits.

In 1856 William Henry Perkin, who was an 18-year-old student, discovered that dyes could be made from coal tar. These **synthetic** dyes are used today instead of natural dyes. Why do you think this is?

Experimenting with natural dyes

Which fruits, vegetables and plants make good dyes? Think about fruits or vegetables which stain your hands when you cut them up, stain your clothes if you spill them or colour the water they are cooked in.

Trying out natural dyes

You will need:

- Some plants, vegetables or fruit (these can be fresh, frozen or canned; you will need about 100g of each)
- An old saucepan and lid for each fruit, vegetable or plant
- An old wooden spoon
- A chopping board
- A knife
- A sieve
- A jug and teaspoon
- Fabric pieces (these should be washed first if they are new) or light coloured yarn
- Paper or newspaper
- Rubber gloves
- An overall

Here's what to do:

1. Cut up and/or crush about 100g of each fruit, vegetable or plant.
2. Put the pieces in a saucepan with 750ml water. Soak it overnight if you have time.
3. Bring the pan to the boil and simmer it for at least ten minutes and up to one hour.
4. Sieve out the pieces and keep the coloured water (this is your dye).
5. Soak your fabric or wool in the dye until the colour is taken up (at least 15 minutes).
6. Remove your fabric or wool from the dye and dry on paper.

Present your results. You could draw up a chart to show the colours you got from different fruits or vegetables.

- Which fruit or vegetable gave the strongest colour?
- Which gave the weakest colour? Why was this?
- Why do you think natural dyes are not often used today?

Mordants

Some fabrics do not dye very easily – the colours are weak and wash out. To help **'fix'** the dye you use a **mordant**. A mordant **chemically** fixes itself to the dye, making it **fast**. Mordants might be salt, vinegar, copper filings, urine, wood ash, or **alum** with cream of tartar.

Ways of colouring fabrics for you to try

Tie and dye

Tie and dye is a very old method of making a pattern on fabric.

You bind parts of the fabric tightly with string or raffia, sometimes around hard objects such as pebbles, beads or rice. When you put the fabric into a bath of dye these tied areas will resist taking up the die and so stay the original colour of the cloth. This is called **resist dyeing**.

Tie and dye from Nigeria

Tritik

Tritik uses hand and machine embroidery to create the patterns. You use buttonhole thread or double sewing thread to make stitches very close together so they will resist the dye. You unpick the stitches after dyeing to reveal the pattern.

Tritik dyeing

Bandhani

In Bandhani you make hundreds of very small ties on the fabric. After dyeing, these ties leave a very intricate dotted pattern.

Bandhani

COLOURING TEXTILES

Ways of colouring fabrics for you to try (continued)

Shibori

This is the Japanese version of tie and dye. You use tiny ties together with carved blocks to give a softer outline.

Indigo dyeing

The Yoruba and Hausa people from Nigeria use a combination of tying, pleating and sewing with indigo dyes. They produce beautiful dark blue fabrics.

Adire eleko

Traditionally, dyers use **cassava** paste to resist the dye, but you can use wheat flour instead. Mix flour and water into a **batter** and put it into an empty, clean washing-up liquid bottle. You can squeeze your pattern on to the fabric, and when your batter has dried you can paint the fabric with the dye.

Shibori

Adire eleko

Indigo dyeing

Ways of colouring fabrics for you to try (continued)

Batik

Batik is another very traditional method of colouring fabric, using wax to resist the dye. Archaeologists have found very old pieces of batik cloth in Egyptian tombs dating back to the first century AD. You can use batik to give very fine decorative work, or large bold patterns.

Batik

Block printing

Block printing is a simple but effective way of decorating fabric. People practise it across the world.

Today it has mainly been replaced by large scale machine printing, except in India where the traditional methods are still used.

You may have seen some of the most famous plant and bird designs, produced in England by William Morris at the end of the 19th century using block printing.

William Morris, *The Strawberry Thief*, 1883

Block printing

FRUIT DESSERT

Layer by layer

Your challenge!

Supermarkets sell a lot of chilled desserts such as yoghurts, trifles, cheesecakes and mousses. Some are the same texture all the way through, others have a variety of textures – crunchy, smooth, chunky, soft, chewy. New types are tried on consumers frequently but some do not sell for long.

Food manufacturers carry out a great deal of research to find out what consumers like. Recent research shows that some people like small, individual layered desserts with fresh fruit in them. They are trying to develop a new product to meet this demand.

Your challenge is to design a layered, chilled dessert containing fruit, for one person. You may present it in a vacuum-formed package which shows off the layered effect.

Why this activity is useful

◆ It is all about developing new products.

◆ You will learn some food science such as how different ingredients can cause different effects like thickening and setting.

◆ You will understand about different types of fruit, how they can be prepared for eating in interesting ways and why they are important in everyone's diet.

◆ You will experiment with different textures, flavours and colours and how they can be combined to make a new dessert.

The broader picture...

Forty years ago there were very few ready-to-eat desserts in shops. What are the disadvantages and advantages of their availability now?

Nowadays, very few new food products sell for more than one or two years. Can you suggest why this is?

'Light' or 'lite' desserts contain less calories or fat. Find out what is used in these desserts to keep them as close as possible to the original recipe. Are 'lite' desserts a healthy alternative to fresh fruit?

To be successful

★ Your research into existing products, your ideas and experiments with ingredients will result in a dessert which contains fruit, combines different layers, colours and textures, looks attractive, and tastes wonderful!

★ You will apply your understanding of how different ingredients may be used to set and thicken food products to create a popular dessert.

★ You will evaluate your product and consult other people appropriately.

★ Any packaging you make will be attractive and complement the appearance of the dessert, keeping it free from contamination.

Planning things through

▸ Keep a detailed record of what you do, beginning with a plan of how you will use your time.

▸ Research into the types of chilled desserts in supermarkets will give you ideas about what is already produced. This should be done for homework.

▸ Allow time for carrying out food science experiments with thickening and setting agents. Also leave time to test and improve your ideas.

Product evaluation – examining and testing desserts

There are so many desserts on the market that it is difficult to think of an original idea. It will help if you make a database of what is already available so you can discover where there are gaps.

Designing and Making: Ideas from existing designs → 94

You will also need to find out what flavours, textures and fruits other people like – the 'features' of the dessert. You could do this by carrying out some market research. Try asking different people questions and recording their responses.

The findings of your market research could be recorded as a star diagram, then you can easily see what the most popular features are.

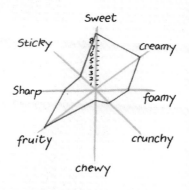

'Tastes, colours and textures people like in their desserts'

PRODUCT EVALUATION : LAYERED DESSERTS			
Manufacturer		Type of fruit used	
Name of Product		How it is made	
Pack size(s)		Special dietary requirements (light, vegan, etc.)	
Main Ingredients		Who is likely to buy this product	

Generating ideas for the dessert

Manufacturers use different ways of thinking up ideas for new products such as sketching and drawing, brainstorming, 'fantastic' ideas, 'blue sky' designing.

Blue-sky designing – no limits on the ideas

From your market research you should have found out:

■ what kinds of desserts are in the shops

■ the features people like in their desserts.

This should help you to begin to record some ideas.

Sketch or note some ideas that you think will be popular. Keep checking back to your market research, especially the star diagram.

Show your designs cut in half so the different layers can be seen. Remember to label and describe each part.

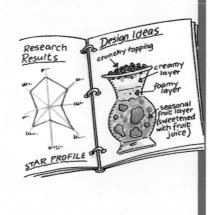

Creating different effects

You will need to carry out some investigations to help you understand the properties of ingredients and how you can use them to create your special dessert. This is food science and will be useful to you when you design and make any food product.

➡ *Designing and Making: Thinking about materials* 104

Foams

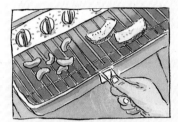

Browning and crisping

Thickening

Gelling

Using fruits

Writing a specification for your design

Choose the best of your ideas to go ahead with. Write a specification to show what the new layered fruit dessert will be like. You should include:

■ what each layer looks like – colours, textures

■ what each layer will taste like

■ what kind(s) of fruit will be used and how these will be prepared

■ who this dessert will appeal to

■ quantities of the various ingredients

■ total quantity (portion size).

➡ *Designing and Making: Specifications* 92

Specification – layered fruit dessert

SPECIFICATION FOR LAYERED FRUIT DESSERT
- creamy custard, light pink, slices of strawberry
- crunchy biscuit using chocolate biscuits
- soft pureed strawberries

Reasons why I chose this idea – my research showed that strawberries were the most popular fruit. People wanted creamy

Trying out your idea

When you have written your specification write out the recipe for the dessert. Then, at home, work out the steps you need to go through to make the dessert. Draw a flow chart to show this process. Remember to include time for the layers to cook, cool or set.

Designing and Making: Planning manufacture 118

Does it pass the test?

Consumers must enjoy eating your dessert or it will not sell successfully. Evaluate your result for yourself and ask other people to taste it.

Compare the findings shown in your star diagram with your dessert. Does your dessert have the features people said they wanted?

The star diagram may help you change or modify your design.

'This is how I'm trying to get my dessert to be.'

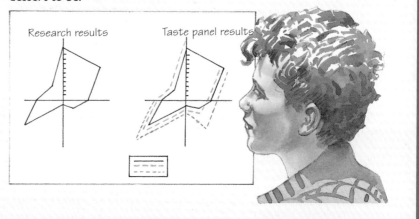

Research results   Taste panel results

Designing a complimentary package

Food packaging is very important. Packages must look good to attract buyers, but they must also protect the food inside, stack safely and fit neatly on supermarket shelves.

Designing a complementary package (continued)

You can design and make your own container for your fruit dessert. Firstly you will need to work out how much volume your container will hold. (How big is your dessert portion?)

For a cylinder use the formula:

$\pi r^2 \times \text{height}$

(ie 3.142 times the radius multiplied by itself, times the height of the container). For a cone do the same but multiply the whole equation by $\frac{1}{3}$.

Are there other ways to make a good estimate of the volume? With a layered dessert you may want to design a container which is also layered. You can make a former and vacuum-form your container using food-safe plastic.

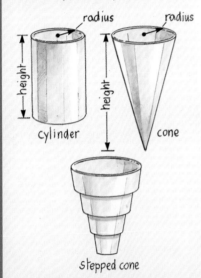

→ Red book: Vacuum-forming   110

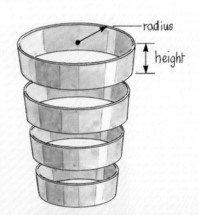

When evaluating your container you will need to ask how the design of the package has helped the visual appeal of the dessert.

Each segment of this stepped cone is a different size of cylinder

Case study – De Roma's twin-pot ice cream

DE ROMA ICE CREAM LTD
Specialist Manufacturers of Quality Choc Ices, Filled Cones and Lollies
Est. 1922

One of the Directors of De Roma Ice Cream, Steven Wetherby, describes the background for the development of their twin-pot dessert.

The ice cream market for 'Cups and Tubs' is worth £3 million each year in England, and has annual sales of 1.4 million litres. Concept research showed that amongst dieters and maintainers, there was a large demand for a reduced-calorie twin-pot product.

The company tried out the idea to see if it would be feasible. A part of this was to make the product on a small scale and try it out on consumers. This is called a pilot-plant trial. We made 100 samples. We worked out how much production would cost and what the nutritional content would be. We asked our designers to design a product to meet the following specification.

Product requirements
It should be like twin-pot yoghurt with ice cream and topping separate. The ice cream could be non-dairy or a frozen dessert, frozen yoghurt or fromage frais. The topping could be spoonable liquid or nuts/crispy textured pieces.

Nutritional requirements
Calories must not exceed 150 Kcals per twin-pot, preferably under 100 Kcals. Energy from fat less than 30% total energy. Free from artificial colours, flavours or preservatives. No artificial sweeteners.

Product size
160 ml, packaged in existing plastic twin-pot with foil or film seal.

Shelf life
12 months minimum, robust enough so that quality is not to be lost in distribution.

What ideas would you have had if you were one of the product development team in this company?

How would you have tested your product?

Remember us

Your challenge!

Souvenirs and collectables are big business. They are used to promote events, pop stars, cartoon characters, fast food outlets, even schools. Products such as T-shirts, mugs, posters and models of characters make a lot of money for the promoters and the people who make them.

You will work in a team. **Your challenge** is to design and make a co-ordinated range of promotional products for a special occasion or a client. You must decide what occasion or who the client is as your first task.

Your team should produce at least six different products using a range of materials. Your Challenge also includes presenting your work as a display.

Why this activity is useful

- You will develop team-working skills, collaborating with others and taking responsibility for your own parts.
- It will give you a chance to develop skills and knowledge from earlier D&T projects.
- You will learn how to create a proposal that must convince a client.
- Their 'image' is very important to a commercial organisation and you will come to understand why this is so.
- You will develop communication skills when presenting and explaining your team's concept in a display.

The broader picture ...

- Why do companies give away promotional items?
- Why do people buy promotional items?
- Are 'free gifts' really free?
- Not everyone is allowed to manufacture products with Mickey Mouse on them. Why is this the case?
- Why do some people complain when new promotional items are brought out, e.g. new football club strips?

To be successful

★ Your team will produce an eye-catching display that communicates:
what or who you are promoting,
the corporate image you have developed,
the range of products you have developed,
the reasons for your group's decisions.

★ You will have designed and made a high quality example of one of the products.

★ You will have thorough records of your ideas, plans, and how you used your time.

★ Your team will have developed a range of products using different materials, and where possible, used CAD/CAM.

Planning things through

With your group, you will be in charge of this Challenge, though your teacher will advise you.

Your team will need to:
use time outside of lessons to find a client and research thoroughly what they require,
decide what products to include in the range,
decide and agree on the corporate image,
organise and prepare your display,
allocate responsibilities to the whole team,
communicate with each other well, throughout the project,
involve your client in evaluating your ideas and work.

What are promotional products?

Look at the range of promotional products on these pages. As a group, list all the different promotional items and collectables that you have seen. Someone should organise them into types (or categories).

What makes a good promotional product? Someone else could list these features.

Designing and Making: Finding more information **91**

Once you have chosen your client or event, brainstorm ideas for your range of products. Discuss each idea and decide which are possible for you to carry out.

Where else to find help

The pictures will remind you of some of the Challenges that you may have already worked on and tell you where you can find help. It will also remind you about the skills, materials and knowledge that you could use in this new task.

Make a spider diagram of ideas, skills and materials that you will use to make the products on your list of ideas.

Red book: Researching information **89**

More about corporate identity

What does 'corporate' mean? Someone in your group should look it up (use more than one dictionary).

What is a company trying to do by having a corporate image?

This might be something to discuss as a team.

You will not be able to develop an appropriate image unless you understand very clearly what you want it to say. What messages does your client want to project? How are you going to find this out? Who in the team will do this? Read the case study below before you go ahead.

Case study – Maynard Leigh Associates

Andy Leigh's story about the development of the corporate image for Maynard Leigh Associates.

Andy Leigh, who had worked in management and Michael Maynard, an actor, founded a new business management training company in 1989.

They wanted to have a brochure, a logo, advertisements and business stationery such as their own notepaper. With the help of a Government grant they employed a designer.

The most important thing the designer did was to talk to them about their company. He asked questions such as:

What is the company really about?
What will you do for your clients?
What is special about your company in comparison with other similar companies?

The designer was trying to find out very clearly what had to be communicated to other people about the company, about its work and its values. The experience also helped Andy and Michael to be clear themselves about what was special that their new company had to offer.

The designer went away and came back with a series of mood boards, some of which were wonderful, but none of them seemed right to Andy and Michael.

What was not being communicated were the unique features of the company – the use of acting and the performing arts to help people to develop management and presentation skills very quickly.

'Unlocking people's potential'. This phrase emerged as the company's mission statement and it was this that inspired the designer to design the perfect company logo. This combines a keyhole (unlocking) and a person (people) of either gender, drawn in a soft, non-threatening style.

Colours were chosen to be vibrant and make people feel good. High standards and high quality were shown in the quality of the paper and high standards of presentation.

Maynard Leigh Associates are very conscious of the value of good design and continue to spend a great deal of money to ensure that their corporate identity is being communicated properly to their clients and people who work in the company.

The images

Your products must be co-ordinated. They must clearly communicate an image or message about your client. Simple logos with simple images are one way. Simplifying images which arise from your research is another.

After you have decided what ideas your images must convey start sketching ways you might do it. Then you will need to discuss which images have the best potential and work on these some more.

Look at this well established logo (above) and how carefully it was changed to make it feel more up-to-date, without losing its familiarity.

The logo (below) implies speed on railway lines and the arrows indicate that trains travel in either direction.

London Transport Property

Computer Aided Designing (CAD) and Computer Aided Manufacturing (CAM)

Creating the right image

When your team have decided on an image to develop, you could use a drawing package or the software of your school's CAD/CAM machines. This will allow you to produce the same high quality shape or image on the different products. Have you an expert in your team who could take this on?

Developing your design

Check that the images you choose will work with the different materials and CAM machines. You may need to adapt them for different materials.

Complete designs will take a long time to manufacture so aim to say as much as possible, as simply as possible.

Designing and Making: Planning manufacture 118

Designing and making

This second part of the book will help you develop the designing and making skills you need to carry out the assignments in the first part. It looks closely at various aspects of designing and making, which are developed further in the other *Challenges* books.

You may use some of these pages with your teacher when learning more about designing and making.

Remember – you should *always* turn to this section *before asking your teacher for help* when you are designing or making and are not sure what to do next.

One of the most important things you should be learning in Design and Technology is to work independently – having your **own** ideas, developing them in your **own** way, understanding for yourself **what** you are doing, and **why** you are doing it that particular way.

Contents

Designing	**90**
Identifying needs	90
Finding more information	91
Specifications	92
Ideas from existing designs	94
Developing ideas	95
Things to think about when designing	
Function	97
Safety	99
Costs	100
Human factors	103
Materials	104
Production	104
Modelling your design ideas	108
Working in teams	109
Presenting design ideas	111
Evaluating	115
Making	**118**
Planning manufacture	118
Making products in quantity	120
Quality in production	123

Designing

Identifying needs

Most of your work in Design and Technology so far is likely to have been doing things you wanted to do or designing for yourself.

Designing something for yourself is always easier than designing for someone else – you know what you want and you know what you like. If you are designing for someone else the first thing you have to find out is what their needs are and what they like. This isn't too difficult if the person is a bit like you.

Professional designers usually have to design products for people and companies they know nothing about, so their first job is to look at the situation where the product will be used. They talk to a variety of people: the people who will be buying the product and the people who will be using it. They talk to the company which will manufacture the product and the company that is paying them to design it (not always the same company).

You should think about questions like: Who is the most important of these people? Will a product that does not work well or which is not attractive to the buyer sell well? How can you as a designer know what your user will want, and need?

For you this means going away from your lessons knowing who you need to talk to, where your product might be used, and organising your time well to research all this information. This is the most important type of homework for design and technology lessons – and it is up to you to organise it!

As you get more information you will have to think which of the needs you identify are the most important.

Floating ideas

Floating ideas means sharing very early thinking with other people to get their responses. Here is an example from one company which shows their way of finding out whether their early ideas are what people want and whether the company's ideas are going in the right direction.

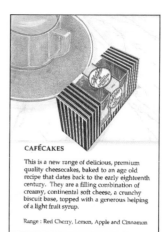

These mood boards were used by Eden Vale. They show two imaginary dairy products which were models of early ideas. As part of their designing they asked customers to look at these pictures and give their opinions of them. This helped them to find out what people liked and were willing to buy. The pictures gave lots of information about the products without having to make real ones.

> Red book: Working with design briefs 87

Evaluating needs

Look and **talk** – these are the two keys to successfully identifying needs: the needs of users, producers, shops and all others who will come into contact with a product.

They asked the customers:
- Would you buy this product? Why or why not?
- When would you eat it?
- How much would you pay for it?
- How many would you buy?
- Imagine you are eating it
  What would it smell like when you opened the package?
  What would it taste like?
  What would it feel like when you bit into it?

Look at the two pictures again. Answer the questions yourself and then ask someone different to answer them too. This should be someone older and, preferably, outside school.
- Are your answers the same?
- Do you both like the same kind of product?
- Do you both want the same things?
- What does this tell you about designing for other people's needs?

This method and any other enquiries you make will help you develop criteria for the success of your product. It will guide you in drawing up a **specification**.

➤ *Designing and Making: Specifications* 92

Finding more information

Once you have begun to look at user's needs, whether others' or your own, you will need to consider which aspects matter the most and what you need to know more about to proceed.

You will need more information and then more precise information, such as:
- what people can do (their physical capability)
- what things people like and dislike (their preferences)
- the range of different people who will use the product
- where and how the product will be stored
- the shapes and sizes of the space where the product will be stored
- the type of occasion on which the product will be used.

All of these help to identify the **constraints** on your product. These are the more or less fixed limits that you must not go beyond, or the requirements which you absolutely must meet.

To obtain this sort of information you will need to:
- closely watch people doing things
- ask them about how and why they are doing them
- produce sketches, diagrams and photographs
- get measurements for things like:
  how much?
  how often?
  how big?
  how many?
  how long?

All this will give you better understanding of the situation you are designing for. You might well discover that the problems involved are very different from what you first assumed.

If your researching of information is good, it will result in better designs.

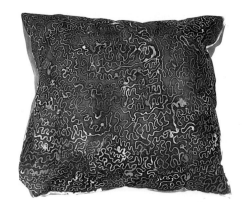

Examples of effective research from students

Specifications

That's what is likely to be said to you if you ever order something and find that what is delivered is wrong!

Specify – what does it mean? As shown in the example (left) it means – 'tell me exactly what you want'.

Try this exercise. In a group, specify what you want from your next pair of shoes. These may be sports shoes, school shoes, cheap shoes or ones to impress others. Each person should write down their own specification for a pair of shoes.

Now look at each others' specifications. Has everyone specified *exactly* what they want?

Identify the categories that have been used and then make sure that everybody in the group specifies what they want in each category.

Above is an example. Have you been exact? Be sure you have been as precise and as detailed as possible, so that if someone went out and bought the shoes for you they would come back with just what you wanted.

This is what happens when designers are working in industry:
1. They might receive a specification that has been written for them and then have to design a product that fits that 'spec'.
2. They may work with others, such as market research people, production managers and financial people to draw up a specification between them.
3. They might be given a general brief and have to work out the detailed specification themselves.

The third way is very common and is shown in the example below.

> Food specification in Fruit Dessert Challenge 81

Notice that specifications state how a product should perform. The detail of how it is to do this comes later. Specifications also state all the constraints that will restrict the possibilities in designing. Some of these will be legal restrictions to do with safety, environmental concerns, etc. They may guide the designer on any of the *Thinking about . . .* sections that follow in this book.

Specifications almost always include measurements. This is important if they are to be exact. If a bridge is to be designed, then the load it has to carry must be specified in tonnes. If a torch is being designed for children, then the ages of the children it is for must be specified in years. This allows the designers to check whether their design meets the specification and does what was expected from it.

For this reason specifications are important when design ideas are being tested and when the products are being evaluated.

> Designing and Making: Evaluating against your specification 115

A specification (left) from the industrial design consultancy Product First and the product it led to

Ideas from existing designs

Starting from a specification, once you know what your product is meant to do, you should be able to produce some ideas straight away. However, the same specification can result in very different products. Here are two camping stoves for similar purposes that are quite different from each other.

Whether some ideas are coming into your head or not, a valuable way to stimulate your ideas is to look at products which others have designed that would meet a similar need.

It is very rare for a designer to produce an idea that is totally original, even for an uncomplicated product. Very often they are being asked to re-design existing products.

It is also rare for products not to have any competition. Someone somewhere will make a product for needs that are at least similar to the ones you are designing for.

Identifying the attributes of existing designs is a good way to understand why products come to be as they are.

> *Red book:*
> *Looking at existing products* **94**

If you criticise other designs it will help you decide what your product will *not* be like. And, you can incorporate the *good* qualities of existing products into your designs. But you are very unlikely to want to use them just as they are – they will need adjusting to meet your specification and to fit with other aspects of your design. Either way your thinking will be clearer if you examine the right existing products.

Remember, while designing
- look at two or three other products to explore their suitability
- ask yourself what is interesting or exciting about them
- look for useable bits of ideas not just the whole thing
- if possible, test them and record your findings as an attribute analysis
- use the information you gather to improve your specification
- evaluate the attributes of your own ideas just as you do others.

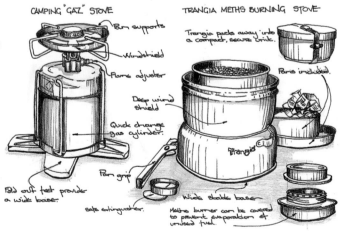

You must expect other people to judge your designs at least in part on whether you have been creative. Looking at others' products while you are designing should help you be clear how your designs will be different and in what ways they will be better than others.

One thing you should always try to do is to meet any special needs your product's user will have. If these are identified in your specification, and if they are made clearer as you look at other products and develop your designs, you will produce something original and worthwhile.

Developing ideas

Being creative and having ideas is an important part of designing, but working on these ideas to develop them is equally important.

Sometimes a designer will start with a range of different ideas, perhaps working on several until one emerges with the most potential. Even then a great deal of work will be involved in developing this idea to meet all of the specification.

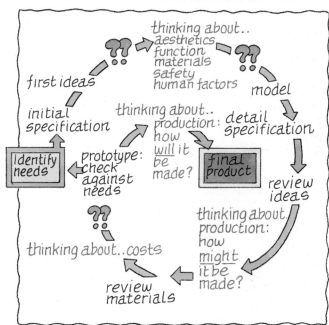

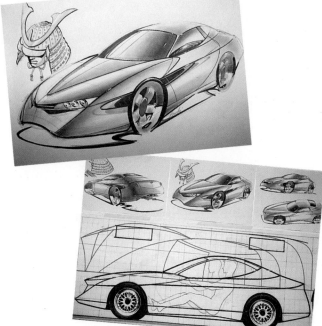

Closing in on a final design

Different types of product will need to go through different stages in their development. Food products will generally take you a shorter time to develop because ideas can be made as finished products, tested quickly, changed and tried again until the best possible outcome is reached.

The more complex the specification, the less the chance that one idea will be good enough straight away. To develop a product which meets its specification may involve a whole series of stages.

However, industrial development of food products does not finish here. Food technologists would then need to develop the product to determine the sell-by date and other legally required information. Also the basic recipe would need to be developed for production in quantity, for example, by cutting back the amount of oil used, using less or more flavouring, etc.

Other products are developed through the making of mock-ups, or other test versions. Sometimes these are used to see if the overall form of the product is right and to find out how easy or difficult it will be to make.

A cheap fabric being used to make a mock-up of a textile product

At other times, it is just one part of a product that needs looking at more closely and testing to see if it works (whether it **functions**).

Whether something works is only one question that has to be answered. **Safety** is another – will the product be safe and not damage people's health? Designing every part of the product to be as economical as possible also matters. If large numbers of the product are to be made then any **cost** saved will be saved a large number of times.

What else might the designer have to think about? Function, safety, cost, and many more factors have to be considered. Use the next

Food product development stages

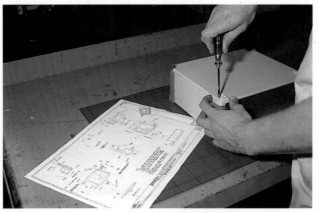

Here a corner protector has been designed and is being fitted to a prototype to test how well the design idea works

section of this book to make sure that you are thinking about enough of the factors that will make your's a good design.

The important thing to do when you are developing your designs is to keep asking questions of your ideas, such as:
- Is this the best way I could do this?
- What alternatives are there?
- Why is this the best method?

You can ask these questions about the whole design idea, or about every part of the product. Remember also to keep checking back to your specification and the Challenge to make sure that you are keeping on track.

Things to think about when designing

Thinking about function

Most products have a function, a job to do. Toasters toast bread, drills drill holes, curtains make a room look attractive, keep out cold air and curious people. When we consider function in design we are talking about whether the product works and whether it does the job it is designed to do.

Your aim when designing is not just to get your product to work, but to get it to work really well. It may have to work in different or changing conditions. Something that works well when it's dry or cold may not work so well when it's wet or hot.

Most products have to do their job without breaking down over a long period of time. **Reliability** and **durability** are important qualities in a well-designed product.

This climber wants to know that her equipment is reliable every time

How do designers make sure that their products are good enough to be used again and again?

Lasting well in use is called **durability**. Are there tests you could carry out to be sure that your product will last an appropriate time?

Food products may require their **shelf-life** and **portion size** to be considered. For example, a fruit dessert packaged in individual portions does not need to last long once opened as the whole amount is eaten at once. Pasta, though, may not all be eaten at once. It is sold in dried form so that what is not eaten at first is preserved. Alternatively, fresh pasta may be specified for freezing.

How do these considerations concern you with the product you are designing? Are you sure you know what the functional demands on it will be? Do you have to consider strength, reliability, durability or shelf-life? If so, how might you test your ideas before you commit yourself?

Case study

IKEA

When someone purchases a piece of furniture they could be spending a lot of money. It is important for them to know that it will last an appropriate time, but this is difficult to determine in the shop.

The Swedish company IKEA use the national Swedish Furniture Research Institute to test their designs under the 'Möbelfakta' system. Methods for testing different features of furniture have been drawn up which vary to suit how the furniture is to be used. Furniture for schools and offices is tested to survive being used 100,000 times. Chairs and sofas for large families or restaurants should have a lifetime of around 50,000 sittings and those for the normal home around 25,000.

Making sure furniture lasts

The Möbelfakta quality standards start with the measurements of the furniture and safety considerations then go on to test
- function,
- choice of materials,
- strength and durability,
- surface resistance (to stand up to spills scratches and mess),
- workmanship.

Chairs, armchairs, sofas, beds, tables, storage furniture, kitchen units, high chairs, cots, office and nursery furniture are all tested before being put on sale.

Thinking about safety

It is the designer's responsibility to make sure that any product is safe to use. To guide them, national and international standards have been drawn up. Some of these say how something must be done, for example, the amount of ventilation in a room is set by building regulations, which have the force of law.

Designers also have standard tests to ensure that their products are safe.

You should always think about the safety of the users of any product you design. Ask some safety questions about your design ideas. Here are some which apply to various projects in this book.

Will it be:
◆ strong enough to be safe?
◆ too heavy?
◆ heavy enough to be stable?
◆ cooked for the right length of time?
◆ kept at the right temperature?

Could people:
◆ trap their fingers?
◆ cut themselves?
◆ poison themselves while making it?
◆ catch an infection from it?
◆ get food poisoning if the cooking instructions are not followed properly?
◆ touch live electricity sources?
◆ catch themselves on anything?
◆ burn themselves?
◆ be allergic to its ingredients?

When developing food product designs, safety is so important it is governed by Acts of Parliament, especially the 1990 Food Safety Act.

Case study

The Jointed Test Finger

Andrew Waddington, Industrial Designer for Vero Electronics Ltd told us how his company tests their designs.

To ensure that their products, which contain high voltage circuits or moving parts, do not present a hazard to their customers, the company tests their products using a jointed test finger. While it may not resemble a human finger very closely, it is designed to mimic the way that a human finger might get into equipment through holes or slots. The finger is an internationally recognised piece of test equipment and its dimensions are precisely defined in European standard BS EN 60529:1992. Its two joints allow it to reach around corners or into areas which are hidden from view, just as a human finger might.

John Parkin, Vero Electronics' Chief Test Engineer testing a circuit board enclosure with the European Standard finger

Many electronic devices require cooling to prevent them from overheating. This is often done by using electric fans which rotate at high speed, presenting a safety risk. Guards are placed around the fans, but these guards must contain holes to allow air through the fan. To ensure that the holes in the guard will not allow a human finger to reach the fans, a European standard finger is pushed in. If it makes contact with the fan, then the fan is not safe to use. The same test can be applied to electronic circuitry which may be dangerous if touched.

Thinking about costs

The materials you use in your designs will affect the **costs** of making products. There are often alternative materials available which may cost less. This becomes very important when products are manufactured in large numbers.

You have to ask yourself questions about costs:
- Are there alternative materials, components or ingredients?
- Are there cheaper materials of the same quality as higher priced ones?
- Could the product be smaller or lighter?
- Are there any unnecessary (redundant) parts?
- Can any component be replaced with smaller, lighter or cheaper ones which do the job just as well?

One way to be more economical is to make more parts from the same material or ingredient. If you can make two parts of a dish from the same flour rather than using some self-raising and some plain flour you will save money. If all, or many, of the parts of a product can be made of the same thickness of the same material you might save money. Why?

Costing materials

Materials are priced by weight, by volume, by length, by area or by the piece.

Priced by weight

Priced by volume

Priced by length

Priced by area

Priced by the piece

First, take each type of material or ingredient in turn and add up the amounts used, eg the total length of wood which is 25 mm × 50 mm in cross-section, or the total amount of milk used. Find out the cost of each of them and add them together for the total cost.

If you need to work out the total area of any sheet material read the section below to make sure you use as little as possible.

You probably will not have to worry about the cost of small 'consumables' such as glue, water, cotton, solder or glasspaper but you should remember that all these have to be paid for. For any small item you may just estimate the cost, eg six small blocks of balsa wood at 5p each.

Here is how a bill of parts for some torch designs might look.

DESIGN 1.

Quantity	Description	Unit Price	Sub total
½ sheet	Polystyrene	30p/sheet	15
Off cut	Acrylic lens		5
2	AA batteries	90p/pair	90
1	Bulb		35
200mm	Wire	20p/metre	4
6	Blocks of MDF	5p each	30
		Total	£1 — 79

DESIGN 2.

Quantity	Description	Unit Price	Sub total
200 mm	Aluminium tube	£6/metre	£1 — 20
Off cut	MDF end cap		20
4	batteries		£2 — 20
30 mm			

Saving materials

We all have a responsibility not to waste materials. Think ahead before cutting out materials from larger pieces, especially sheets of wood and plastics or fabrics. Is there any way you can do it with less waste or taking less time?

Some companies make garments in thousands. They use computers to 'lay' the pattern on the fabric with the least possible waste.

Food products may be made to lower standards for some brands than others. Top brand chocolate biscuits may have a thicker chocolate coating than others but this must be reflected in their price.

One of the problems for you is that the materials for your products may cost you more than a similar finished item costs in a shop, e.g. ballpoint pens only cost about 15p. But think what is in them.

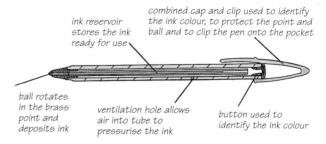

Why are products which are made in large numbers so much cheaper?

Fitting shapes together to waste the least possible is called **nesting**. In materials like plywood or fabrics this is a very important way to save waste. Sheets of clay or food materials, such as pastry can be nested uneconomically as long as you re-combine the waste, roll it out again and use it.

The costs of batches of products

Saving material is even more important if you are making more than one item. You must plan even more carefully to avoid waste by, for example, fitting together pieces you want to cut out.

Planning food ingredients for economy

Often there are a lot of ingredients in food recipes and more of one ingredient might mean more or less of the others. The best way to plan this is to use a computer spreadsheet.

Carefully nested 'lays' of garment pattern pieces on fabric, one worked out with a computer program

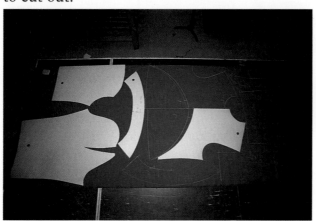

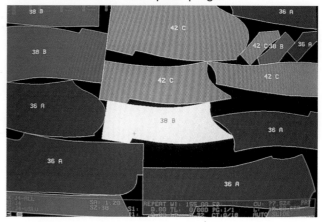

These students carried out taste tests first and then produced recipes based on these for further tests on people who represent their identified market. Here they are checking the nutritional value as well as the costings for producing a batch. Costs will vary depending on the number of products produced. The spreadsheet allows them to model costs for different amounts.

Thinking about 'human factors'

Designers have to consider how their products will be used and who will be using them. This is particularly important at the product development stage. Some people may find it difficult to use products which are too big or too heavy, others may find that they are unable to use them at all. Products which are too complex might easily be misused making them less safe.

Changes can be made to make products more 'people-friendly'. Bicycles used to have the gear lever on the frame and cyclists had to look down to change gear. Sometimes they fell off or hit something as a result. Then the changer levers were put on the handlebar, and now many of them just need a twist of the hand-grip to change gear. This is obviously safer because it allows cyclists to concentrate on the road.

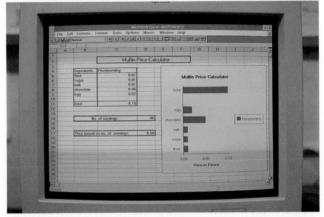

Students at St Thomas School, Exeter using spreadsheets for recipe costings

Three generations of bicycle gear change – a good example of design development

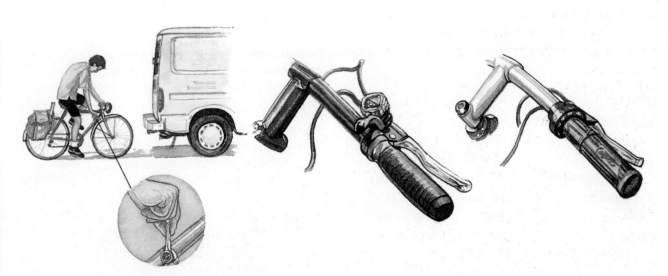

The problem with people is that they are all sorts of shapes and sizes. Design something to be the right size for one person and it may be quite wrong for someone else. This is why it is important to know who you are designing for. If you are designing a product for people of different shapes and sizes, you have to think about how it can be made so they can all use it comfortably.

> Designing and Making: Identifying needs — 90

You may need to look up the measurements for a range of people who will use your product. Anthropometric tables can be used for this.

> Designing to suit your client, Torch Modelling Challenge — 41

If your client is one particular person, or if they have special needs, then tables of information will not be useful. You have to go to your client and assess them for yourself, perhaps taking measurements from them.

Studying how people and products work together is known as **ergonomics**. It is very important to manufacturing industry.

> *Ergonomics as it relates to the design of motor vehicles is the consideration of human factors in the efficient layout of controls in the driver environment. Today it is a fundamental component in the process of designing cars and commercial vehicles.*
>
> The Ford Motor Company

Thinking about materials

Think about your choice of materials at a very early stage in your designing. Think about alternatives you might use. Look what is available as soon as you have an idea that might be worth developing. Then ask yourself
- does this idea still seem feasible?
- am I using the best material for the job?

(See also costs questions on p 100).

Should you alter your design to accommodate the materials available?

Use the tables on the next three pages to help you decide which materials to use in your products.

Thinking about production

As you develop your design ideas, you need to think about how these can be turned into products – how they will be made. There is no point in designing something which you are incapable of making.

This is one of the constraints on your designing, but this does not mean you should play safe and go for easy options.

Go for your best ideas and ask your teacher and anyone else for advice on how they can be made. Make your own decisions on the basis of this advice and stretch your skills to the limit.

Think through the steps involved in making a particular design before committing yourself to it!

> Designing and Making: Planning manufacture — 118

On some occasions, when you plan the manufacture of batches (large numbers) of products, you will need to give more thought to how your designs can be produced efficiently and economically. Professional designers sometimes have some very difficult constraints placed on them by manufacturing requirements in the specification, as this example shows.

> Product First specification — 93

Thinking about materials

MATERIALS	FEATURES	SOME PROPERTIES IN USE	PRODUCTS USING THESE PROPERTIES
WOODS			
Softwood (Pine or Red Deal)	light colour, long fibres, knots, low density, inexpensive	bends but splits breaks at knots	door frames, skirting boards, country-style furniture
MDF (Medium Density Fibreboard)	man-made, large boards, smooth surface	machines well (eg drill/saw/router), paints well	flat board-based furniture with decorated edges
Ash	long grain, low density	bends easily light colour	Windsor chair backs, curved screens clear finished, sometimes stained
Mahogany	dark colour, rich appearance	very 'workable', often stained darker still	furniture, especially highly decorated, eg carved
METALS			
Mild steel	silver-grey, strong, rusts, bends, machines & welds easily	forms rods, sheets, round & square tubes	furniture, brackets, lamps, car bodies, girders, bridges, filing cabinets, fires
Aluminium	silver-grey, soft, usually alloyed	bends easily, will break, difficult to to weld	saucepans, zips
Brass	yellow colour	attractive when polished conducts electricity well	shiny bath taps, low cost jewellery, connector pins on plugs
Titanium	light weight, stiff, machines well	takes colour & stains well	mountain bike components eg quick-release levers
THERMO-PLASTICS			
Polyethylene (polythene)	low density variety (L-D)	both easily moulded, and re-moulded	plastic bags, food storage containers
	high density variety (H-D)		washing-up bowls, food storage container lids
Polystyrene	high-impact type (HIPs) stiff, light, easily glued	easily vacuum-formed	instrument cases – remote controls/ telephones/Game-boys
THERMO-SET PLASTICS			
Acrylic	stiff, transparent, dyes well	glass-like, shatters	transparent boxes/lids
PTFE (Poly-tetra-fluoro-ethylene)	fairly soft, heat resistant	very low friction	non-stick coatings for saucepans
Polyester	liquid resin sets with catalyst, brittle	reinforced easily eg with glass/kevlar/carbon fibres	canoes, motorcycle panniers, car bodies
SYNTHETIC FIBRES			
Polyester	dyes well	strong, light	sheets/duvet covers (with cotton)
Acrylic	dyes well	strong, light	cheap, light pullovers
PET	light, fluffs well, strong	stretchy, good insulator	fleece jackets, dressing gowns
NATURAL FIBRES			
Cotton	light, cool, dyes very well	strong, thin	shirts/blouses, socks, jeans
Wool	warm, fluffs & dyes well	good insulator	jumpers, blankets, socks
Hemp	smelly, coarse texture	strong, hairy, natural look	rope, macramé twine
Silk	colours well	light, shiny	waistcoats, ties, shirts

Thinking about components

PROPERTIES	COMPONENTS	SOME PROPERTIES IN USE	PRODUCTS USING THESE PROPERTIES
Electronics	resistor	resists flow of current	most electrical circuits
	LDR	changes resistance when receives light	street light switch
	push switch	connects current when pressed	computer monitor
	thermistor	changes resistance with temperature change	automatic cooling fan
	transistor	switches or amplifies signal	amplifiers, radios
	battery	provides electrical energy	personal stereo, torch
	potentiometer	varies a voltage	dimmer light switch
	microphone	outputs when sound waves received	rock band
	bulb	lights up when power is fed to it	torch
	buzzer	buzzes when power is fed to it	oven time-warning
	motor	turns when power is fed into it	personal tape-player
	solenoid	changes electrical to mechanical energy	automatic door lock
Mechanisms	lever	amplifies movement & force	tyre removal lever
	linkage	moves force and motion to where needed	bike brakes
	inclined plane	sloping surface to reduce effort in movement	screw, wheelchair ramp
	belt & pulley	transmit rotary movement	fan motor to blades
	chain and sprocket	transmit rotary movement	pedals to wheel of bike
Pneumatics	pneumatic cylinder	pushes a piston	open a bus door
	port valve	releases flow of air into a cylinder	lorry brakes
	flow control valve	varies the flow of air into cylinder	door opening speed-adjuster
	time delay	delays the action of a valve	tube train door-closer
	shuttle valve	manual emergency control	tube train door-opener
Hydraulics	piston	pushes levers or linkages	car engine
	hydraulic cylinder	pushes a piston	car brakes
	fluid reservoir	stores hydraulic fluid	car jack, mechanical digger

Thinking about ingredients

PROPERTIES	FOODSTUFFS	SOME PROPERTIES IN USE	PRODUCTS USING THESE PROPERTIES
Aeration (physical)	eggs	gives a light foamy texture or consistency	cakes, bases for desserts, soufflés, sauces
	egg whites		meringues, cheesecakes
	evaporated milk		jelly whip, soufflés
	cream		mousses, fruit fools, sweet & savoury sauces
Thickening	flours – plain white/ wholemeal	ingredients are mixed with starch & heated to form a semi-liquid, semi-solid 'gel'	sauces, casseroles
	pasta flour		spaghetti, lasagne
	cornflour		sauces, pie fillings, blancmange
	custard powder		fruit fools, trifles
	pea, bean flour		savoury recipes
	arrowroot		fruit glazes
	commercial products e.g. Quikjel		fruit glaze, limited flavours
	sauce & gravy powders		meat, vegetarian, sweets
	packet soups		wide choice of flavours
	sauces	concentration: evaporate the liquid by boiling rapidly	tomato sauce, raspberry coulis

PROPERTIES	FOODSTUFFS	SOME PROPERTIES IN USE	PRODUCTS USING THESE PROPERTIES
Setting	gelatine or agar – agar for vegetarians added to sweet or savoury foods pectin sold as Certose – useful for getting a better set with some fruits	to give a firm/moist texture, a gel	jellies, mousses, soufflés, cheesecakes, yoghurts jams (eg strawberries)
Coagulation	eggs, milk & cream often combined	gives a solid consistency overheating or excessive acidity causes irreversible changes in solids & liquids	custards, mousses, carbonara sauce
Thinning	milk stock water fruit juices	liquid may be required to make a recipe thinner	sauces sauces drinks, sauces sauces
Bulking	cereals e.g. crushed cornflakes oats pasta vegetables e.g. potatoes, peas & beans apples	ingredients may be used to fill out a recipe because it is too expensive, too intense a flavour or to improve mouth feel	crumble toppings dessert bases small pasta shapes in soups savoury pies, sauces pasties apple & strawberry or blackcurrant pie
Sweetening	sugar sugar substitutes in low calorie products honey helps to keep a moist texture syrup concentrated fruit juice malt & molasses – very strong flavour	sweetening may improve flavour, or mouth feel	fruit desserts, caramel topping or flavour mousses, yoghurts, sauces, jams, gingerbread, flapjack cakes, biscuits fruit jams & spreads cakes & biscuits
Flavouring	fresh & dried herbs vegetables fruit meat spices stock cubes ready made sauces	there are times when a particular flavour is needed, e.g. lemon or we need to make a spicy flavour sometimes we use the real ingredients or chemical alternatives	sauces, salads, pasta mushroom soup, tomato purée, pasta apricot compote, lemon juice or flavouring bacon burgers, bacon flavoured crisps cinnamon buns, vanilla yoghurt Mexican, Italian, chicken horseradish, Tabasco

Modelling your design ideas

To develop your ideas means looking at them over and over again, considering all the things we have suggested need *Thinking about*. Each time you re-consider an idea, thinking about a different aspect of it is like going around a circle again, closing in on the final design.

Developing ideas diagram 95

To consider ideas from different points of view means you must **model** (or represent to yourself) your ideas in different ways. There is a lot of guidance on modelling in the Red students' book. Modelling the materials you may use is likely to need a **mock-up** using real materials (without cutting them) or very realistic drawings. Drawings will almost always need annotations to complete the information they give. Function can be modelled through a **'lash-up'**. Costs will need to be modelled through listings, or preferably a computer spreadsheet.

Using many materials it is possible to model how something will finally look and feel. This can also be useful to test some of the human factors such as ergonomics.

Multi-material modelling

Often a variety of materials can be used to build a mock-up of an idea. These can be materials normally found in school workshops or junk materials which are suitable to represent a part in the model.

This student, after drawing several designs to customise a Mini car, used off-cuts of man-made board and acrylic sheet to produce a relief model. Car body repair fillers are also used to help mould the required shapes. You can buy items such as wheels, lights and door handles from model shops and add them to give more realism.

Torch Modelling Challenge 43

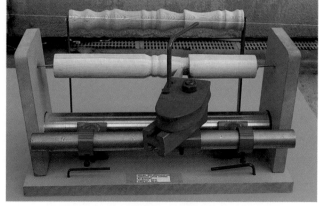

Here a model of a lathe **copier** device has been constructed using man-made board, soft wood, steel tubing, steel rod, and a variety of fittings found in school workshops.

This shows just the basic concept behind the idea so is often called a **concept (or conceptual) model**.

The model of the wind generator has been constructed using cardboard tube for the body, balsa wood for the nose and polystyrene strips for the rotor blades. The gear housing is made from wood dowels, man-made board and polythene tubing.

Wind Generator Challenge 52

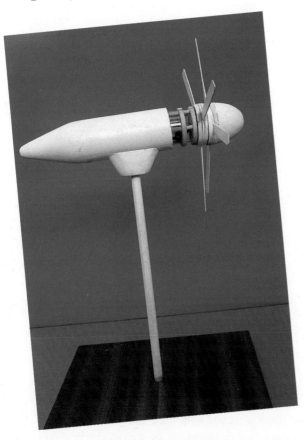

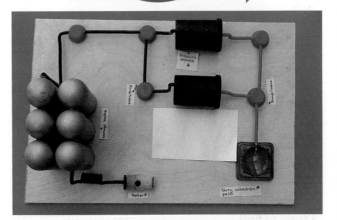

Junk modelling has been used to model an idea for a plant to produce energy from slurry. A whole variety of items have been used including plastic bottle tops, plastic film, cans and empty containers.

Working in teams

Many of the assignments in this book expect you to work in a team. What is the value of working in teams?

Which of the following gives a good description of teamwork?

Picture 1
A group of students are sitting at one table. They are all doing different and/or separate tasks and aren't talking to each other

Picture 2
A group of students are sitting at a table with the same set task, working alongside each other. Occasionally they share notes and help each other with difficult parts of the work. They still spend most of the time working on their own

Picture 3
A group of students have a single, larger set task which they are breaking down into smaller chunks. Each student has a different part of the overall task to complete. They all share ideas and help each other a lot, to make sure that in the end they complete the overall task together

Picture 1: At the same table, working by themselves *Teamwork? – No*

Picture 2: At the same table, same work, but helping each other out *Teamwork? – Not really*

Picture 3: At the same table, one task, with work shared out and brought back together *Teamwork? – Yes!*

When you are working as a team you will be

- sharing the work, ideas, problems and successes
- listening to others to help clarify your own thinking
- building on each others' ideas
- supporting others when they need it, but without interfering
- challenging others' ideas, but learning to trust others' judgements as well
- contributing to the overall objective as well as your own part of it all
- applying your own skills and working hard on your part of the whole venture.

Presenting design ideas

Design folios

A **design folio** is an important record of the work that you do on a project. It will help you as your work progresses and it will help your teacher understand your thinking. It will also be useful for you when you evaluate your work when it's finished and for your teacher to assess it.

Presenting your work to a good standard will give you valuable practice in organising your thoughts and communicating them to others. It will be a chance to develop your graphics skills, handwriting and perhaps IT.

You can use different methods of presentation for the various sections of your folio. You will develop an individual style as you gain experience.

Your folio should be
- an individual record of all work carried out
- presented in a logical order
- bright, cheerful and attractive.

Informal material

Some of your thinking will be shown very informally through initial sketches, collages, notes etc. This allows you to make rapid progress and captures the spirit of your way of working. DO NOT RE-DO IT. You might find that you need to add more notes to this work to make sure that others can understand it. These would clarify details or give some of the reasons to explain ideas and decisions.

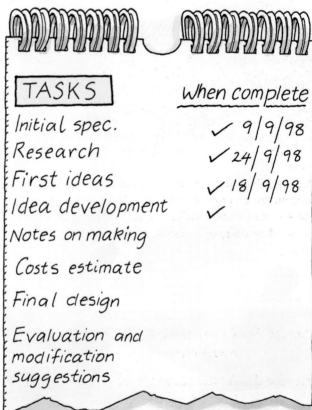

Typical design folio contents (also useful to check against while a project is in progress)

Using bright colours is a good way to highlight the parts you want people to notice and make the sheets more attractive. Overlapping drawings and notes adds excitement and can prevent a sheet from looking empty. You can also cut out small amounts from some sheets and paste them together as a collage of thinking.

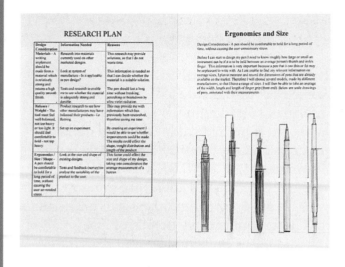

Formal presentation

Written sections of a folio containing such as specifications, research and evaluation tend to be formal so a neater, more careful style of presentation is used. You might use word processing or desktop-publishing (DTP) software on a computer for this.

> Red book: Presenting for maximum impact 83

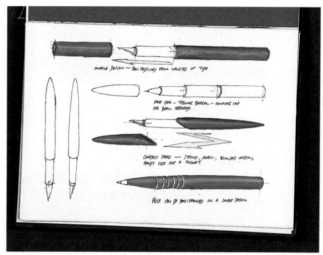

For some parts of the work you will need to use a combination of pictures and text but for presentation purposes you should treat these the same. As you arrange pages of your folio, look carefully at the arrangement of colours and textures. Stand back, blur your eyes and see which parts of a page are heavy and which are light. Be sure to have some blank spaces to counterbalance the heavy parts.

Formal pages, whether text or graphics, should have less information on them, more carefully arranged. Informal pages may be very full and irregularly laid out.

If you have worked in a team you may include work from others but always make very clear which is yours and which is not.

Displays

1 Select the work you will use in the display

2 Choose a background colour and material that compliments your work

3 Cover the background as thoroughly and neatly as possible

4 Arrange the work on the backing, temporarily at first. Stand back to view. Include all labels and 3D pieces. Re-arrange it until satisfied

5 Fix the work firmly

6 Invite your guests!

The impact of displays on the school

If your work is regularly displayed around the school it will help year groups younger than yourself to know what they should aim for. It will also help you to see other people's work.

Collections of similar products can help you see how differently designers have solved the same problem.

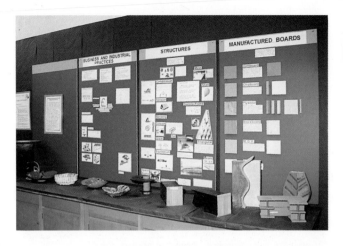

Displays can also be used to raise issues as newspapers do.

Evaluating

Evaluating your ideas
When you start designing you may have a number of possible design ideas.
- How can you decide which ideas are good ones?
- What can you do to help you sort your ideas out?
- How will you choose which ideas are worth developing into fully working designs?

As you develop your designs, you need to evaluate them all the time. You should always be asking questions.

Modelling can help you to evaluate different ideas. Sketches, working models and prototypes can be used for this purpose.

Evaluating against your specification
The purpose of a specification is to ensure that the designer produces products which meet the needs that have been addressed, within the constraints that have been identified. The examples of specifications in this book (see pages 83 and 93) are

Why do it this way? How well does it do it? Is it likely to work well? Is it safe? What are it's strengths and weaknesses? Am I able to achieve it? Does it look good and feel good?

probably a bit more demanding than your's. However, no matter how basic your specification is it will still be useful when you come to evaluate your product, and the way you have worked.

Have I fulfilled my purposes?

This is the basic question for evaluating a product that you have designed. So start by looking back to the beginning of the assignment. Check these sources that told you what was needed:
- the Challenge
- any design brief or other instructions from your teacher
- your specification (and any revisions to it).

Perhaps the best way to evaluate your product is to use it for what it was intended. Examine your product 'with new eyes' – try to look at it as if you have never seen it before. Get others to help you do this, preferably those for whom it was designed.

This student is visiting a nursery school to evaluate the book she has designed and made for young children

User testing is carried out in industry before a new product is launched. As well as physical tests, groups of potential users are asked for their opinions about a new product.

IKEA case study   98

Does it look, feel, taste, smell or sound good?

It might be necessary to change or modify your product after it has been tested.

How to test your product

Which test is best to use for your product? This will depend on what you need to find out about it. If you were testing the following products which test(s) would you pick?

Whatever test you choose you must set it up so that it is a fair test. You must be accurate in recording and presenting the results. Each product or sample must be dealt with in the same way.

	1	2	3	4	5
a					
b					
c					
d					
e					
f					
g					
h					

If you can show that your final evaluation has increased your understanding of the designing and making assignment, then this should improve your marks on that assessment. It should also help you in future work and increase your satisfaction from this project.

Evaluating your work in the team

A team works well when everybody pulls their weight. To help the work of your team it is important to evaluate your own contribution. Often people only evaluate their performance at the end of a Challenge. However if you start thinking about some of the following questions as you go along, it might help you to do a better job in the first place!

Make out a grid like the one here and give yourself a rating on a scale of 1 to 5 for the following questions. '1' corresponds to 'no contribution to the team effort at all' and '5' to being 'fully supportive of the team effort'.

To what extent did you help (are you helping) to:
a break down the Challenge into smaller tasks?
b identify ideas, resources and skills to help solve the Challenge?
c put the tasks into an effective working order?
d allocate tasks appropriately to team members?
e complete your own tasks?
f assist others to complete their tasks?
g bring the whole thing together at the end?
h present the completed Challenge in an effective way?

At the end you could also find it useful to make your own judgements on the performance of the team as a whole. Compare these results for the whole team with your own results.

Finally, identify the part where you made the largest contribution to the team effort. Then identify the part where you feel that you made the least contribution. Make a note of these two for future reference, so you can build on your best contributions and be aware of your weaker areas.

Don't despair – the best teams are those where each member has a different best contribution to make. The worst teams are those where all members focus on the same skill and ignore the rest. Maybe you know some adult teams that behave in these ways?

Making

Planning manufacture

Thinking about making your design

Many design decisions are concerned with how a product looks (**aesthetics**) or how it works (**function**). But some of the more difficult decisions are to do with how the product is going to be made. The list below includes some of the key questions you will need to ask yourself before you start to make your product.

- How many identical products need to be made?
- How much time is available?
- What restrictions are there on cost?
- What materials are available?
- What tools and equipment are available?
- What skills do I have and what new skills might I need to learn?

Decisions will be affected by the quantity of products that you need to make. You would not usually work the same way to produce a hundred products as you would to make a single item.

Much of the work you do will be on your own. Some will involve you working as part of a team. Sometimes you will only need to make a single product. Sometimes you will make a single product but design it in a way that would allow it to be made in quantity. A single product like that is usually called a **prototype**. At other times, you may be required to make lots of products which are identical.

This garment rack is being made by hand as a prototype. If it goes into commercial production, further developments will need to be made

Planning for manufacture

It is a good idea to have a plan which outlines the jobs that need to be done and their order – a simple list or flow diagram will do.

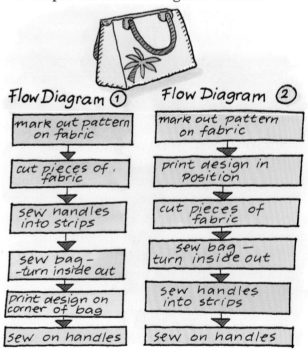

Which order of manufacture would be the best one to follow?

In the past, you may have been able to keep all of your planning in your head. This is possible to do when the products you are making are very simple. However, as your work progresses, planning takes a larger part of your time if you are to avoid major problems. Can you imagine what the planning must have been like for the team who built the space shuttles?

By the time you take your GCSE examination your planning will need to be quite detailed. You will undertake complex projects which involve doing several jobs at the same time.

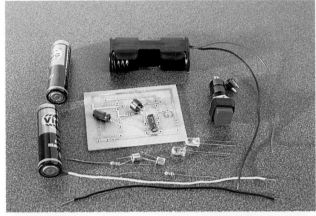

Computers can be used to aid planning

Good planning will ensure that
◆ your time is used well
◆ all your materials and equipment are available at the right time
◆ tasks are undertaken in the best possible order.

So, what does this mean in practice?

When building this circuit, the order of tasks is less important. It does not matter if the LEDs are soldered before or after the resistors. If everyone in the class needs the same circuit, there are lots of ways in which you can work as a production team. By planning out who does what, and by carefully organising your work spaces, you can build the circuits much quicker and more easily than by working on your own.

This is good planning for manufacture.

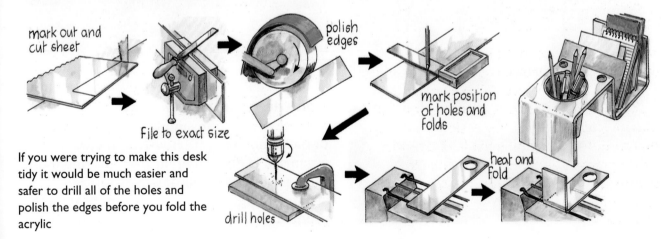

If you were trying to make this desk tidy it would be much easier and safer to drill all of the holes and polish the edges before you fold the acrylic

Making products in quantity

We talk about **one-off production** and **volume (or high volume) production**. Have you ever considered how different these might be?

One-off production

One-off production is used for many different products. Buildings, for instance, are often one-offs although they may be designed to use parts made in quantity elsewhere, such as window frames or bricks. A carnival float, a wedding dress and a visitors' centre display may be other examples.

Unforeseen problems almost always arise during making stages even when the designer is also the maker, as you will already know from your own experiences. Some of these can be avoided through careful planning. With complex products like buildings, good planning is essential.

One-off production is often undertaken by a single person. For example, a confectioner making a wedding cake might do every stage of the cooking and decorating on their own.

Volume production

Volume production simply means making lots of the same product. However, the way it is organised is very different from making the same product once and then again, and again. Also, good planning before beginning production becomes even more important.

To make more products more quickly, various different approaches are used. Here are some of the most common features:
- use a lot of people, or use automated machines
- divide up the making of the product into very simple steps
- give each step to a different person (or machine)
- make it as easy as possible to do each step
- pass the product along a line of workers in the order needed to make it.

By this means the products are made as quickly as possible, and no worker needs to be very skilled, except in the small task each does.

Using jigs

Both in factories and school workshops jigs are often used to help make identical products. A mitre box is an example of a simple jig which is available in shops. It has been made to aid the sawing of 45 degree angles (mitres) for such tasks as making picture frames. It is suitable for cutting small numbers of mitres but a more sophisticated jig is required if lots of mitres are going to be cut. Of course, a fully automated machine designed specifically for this one job might be produced if the quantity justified it.

Cutting lengths of steel

Sharpening one end

Forming the heads

Packing

Crosrol case study 23

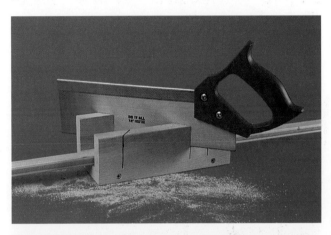

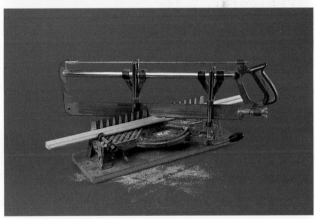

So, a jig is something which aids making by either reducing time, or increasing accuracy, or both. They might also be used to improve safety for a machine operator.

Manufacturing industry makes great use of jigs especially for tasks such as assembling several different parts, drilling holes in set places and sewing accurately.

This jig is used to ensure that all the components are correctly lined up before the seat belt is sewn together

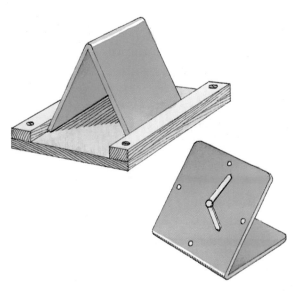

A jig for folding

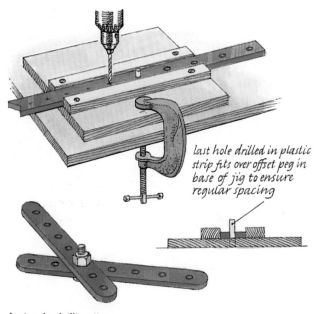

last hole drilled in plastic strip fits over offset peg in base of jig to ensure regular spacing

A simple drilling jig

Sometimes you will need to design and make your own jigs. These might be very simple devices. The drilling jig above could be made from three strips of timber glued and pinned onto a base of chipboard. Once securely clamped to the table of a drilling machine it allows each hole to be accurately positioned without measuring. This jig would only be worth making if you had a number of these holes to drill.

However simple a jig may look, it is vital that it is accurately made otherwise faults will be reproduced on each item being made.

This jig has been made to ensure that each piece of acrylic is folded in the same position. This could be made simply from two pieces of board glued and screwed together with a third piece attached to set the exact position of the fold. Perhaps you could think of ways to make this adjustable so that the same jig could be used for other folding jobs?

Division of labour

One of the best ways of making products in large numbers is to work in teams. Each member of the production team is given a specific task to do, which they repeat over and over again, becoming very fast and skilled at doing it. This is called the division of labour. Some factories divide up tasks between hundreds of workers. For your purposes, a team of four or five would be ideal. Think about:
- Who will pick the team?
- How will they choose?
- How will tasks be allocated?

These are sometimes difficult decisions for employers. They place advertisements in newspapers and hold interviews. It takes a great deal of time and they sometimes make the wrong choices even then!

Decide which people would be best for each task

Here is an example of a team who have produced a batch of greeting cards. They decided that they would all make a copy of the same design and everyone made two cards each. Do you think they made the best decision? How would you organise the tasks between your team?

How can you ensure all of the products in your batch are produced to the same standard?

Quality in production

When planning for production you must consider how you will make sure that your product will be of good quality – 'right first time' is the aim.

In industry it is very expensive if some of the products a company makes are not up to standard. These will be wasted or, if sold, may be returned. No one will buy from a company that becomes known for poor quality products.

Sometimes it is even more serious than that if quality drops.

Case study – Allied Signal

Quality on the road
We spoke to Julie E Pennell a Project Manager at Allied Signal.

At their Carlisle unit they design and manufacture safety restraint systems, such as car seat belts and airbags for a worldwide market. Quality and customer satisfaction go hand-in-hand because every single product that is made may one day save a life. This means that every product has to be perfect. To make sure that no mistakes are made, quality at Allied Signal is a part of life and begins even before a first concept drawing is made.

Throughout the company, activities which affect product quality are carefully controlled.

Allied Signal have certification to the international standard BS EN ISO 9000, a recognised standard which ensures that quality systems are in place.

Designing a product
In some cases it takes several years from concept design to a finished product going into cars for sale to the public. The

Case study – Allied Signal (continued)

designers work closely with the customer, the car manufacturer, to agree exactly what the requirements are. How will it be fitted into the car? What colour should it be? Which country is the car going to be sold in? The product will have to gain legal approval before the safety belt or air bag can be fitted into the car.

Quality built in

Once a design has been agreed they have to make sure that all the materials used will also meet the required standard. Metal parts must be strong enough in a crash, for example, and all the parts must last for the life of the car.

The importance of quality continues from design through to manufacture. All manufacturing processes are carefully designed to ensure high quality. Some machines even have a digital display telling the operator if something is wrong before it is assembled.

Work instructions and procedures are placed on each work station, and all the quality documentation is carefully controlled. Each batch of safety restraints that is produced is given a unique number so they can trace back exactly the details relating to that particular batch.

Each highly skilled operator is trained in quality standards and they each check their work before it is passed on to the next operation. They will reject anything they feel is suspect. Special in-depth checks, called Product Audits, are carried out by specially trained quality technicians.

However, the work is not over once the finished product is packed for despatch to the car manufacturer.

Special testing takes place before the first product is sold and then regular samples are taken from every batch that is made. This is called Conformity of Product Testing. Every product must be perfect and conform to the same standard.

With careful planning, design and testing of the product, coupled with the control of manufacturing processes and the highly skilled workforce, Allied Signal can ensure that its safety restraints will function every time.

A digital display (right) tells the operator that the seat belt is working

Quality in food production

Here are some photographs from Pennine Foods, a large food production company, which show how hygiene and safety are important aspects of quality with food products.

Ingredients are stored carefully in date order and the system controlled with bar codes. What systems will the bar codes control?

The quality and freshness of ingredients are checked thoroughly

Ingredients are stored in the best conditions and temperatures. Why are potatoes kept in dark, cool conditions?

Making

All equipment and surfaces are kept very clean and are regularly sterilised. They have smooth, easily cleaned surfaces. Why is stainless steel often used?

Raw food is prepared in a different part of the production area, well away from any cooked food. Equipment and other materials used in each area are never allowed to get muddled up. Why is this so important?

Everyone who goes into the plant wears sterile protective clothing that covers over most of their skin and own clothes. If they leave the production area, or go from an area with raw food to one with cooked food they have to change their protective clothing.

Sensors are used to detect foreign bodies such as metal in the finished food

Index

3D glasses 8
   modelling 22
   pictures 8
Adire eleko dyeing 76
aeration, food 106
aesthetics 118
anthropometrics 42
applications, components 106
   food processes 106
   materials 105
attributes analysis 40

Bandhani dyeing 75
Batik dyeing 77
bending steel 14
block printing 77
brainstorming 80
BSI kite mark 60
bulking, food 107
buttonholes 61
buttons 62

CAD/CAM 23, 88
case study, Allied Signal 123
   Crosrol 23
   De Roma 83
   Dixon's 20
   Halfords 60
   Heinz 28
   Hinchcliffe & Barber 49
   IKEA 98
   Maynard Leigh 87
   Petzl Zoom Headtorch 42
   Queen Mother's gates 15
   Vero Electronics 100
ceramics, decoration 45
   design project 44
check list 57
circles 10
circuit, electronic 58
clay rolling 48
   slabs 48
   slip 48
client profiling 20
coagulation, food 107
colour complements 72
   contrasts 72
   design project 70
   fixing 74
   wheel 72
colouring fabrics 75
   textiles 72

complementary colours 72
compression 14
computer-aided design 88
constraints, product 91
contrasting colours 72
control of quality 29
   of weight 31
corporate identity 87
cost saving 101
costing 100
cover designs 66
culture portrayal 65
customer's needs 21
cutting steel 16

decoration, textile 63
decorations, ceramics 45
design constraints 91
   client 40
   dessert 81
   development 47
   folio 22, 111
   food products 96
   function 97
   human factors 103
   lettering 68
   mechanisms 67
   rotor blades 52
   safety 99
   specification 21
   teams 69
   textiles 96
   wind generator 53
designing 90
dessert design 81
project 78
dipping, colour 70
displays 113
door stop 15
drawing fish 46
dribbling, colour 70
durability 98
dye fixing 74
dyeing, colour 70, 75
dyes 73

economic design 102
edging, textile 61
electricity generation 54
electronic circuit design 58
   textiles 58, 60

electronics, properties 106
energy conversion 54
ergonomics 42
evaluating 37, 49, 115, 117

fabric colouring 75
   printing 77
fastenings, textile 62
fibre properties 105
finish, steel 17
finishing, textile 61
fish dish project 44
   drawing 46
   model design 45
fixing dye colours 74
flashing lights project 56
flavouring, food 107
floating ideas 90
folding steel 18
food packaging 82
   product design 96
   quality 125
formal presentation 112
forming metal 13
fruit dessert project 80
functional design 97, 118

gear wheels 55

hems, textile 61
human factors 103
hydraulics, properties 106

ideas development 95
   generation 80, 90, 94
illusion graphics 9
image communication 88
indigo dyeing 76
information gathering 91
Italian food project 24

jigs 121
joining steel 18
   textiles 61

labour division 122
large scale production 29, 120
LED circuit 58
LEDs, flashing 56, 58
lettering, design 68
lines 10
logos 88

making   89, 118
manufacturing plan   23
market analysis   41
markets, teenage   20
material costs   100
   properties   105
materials   104
mechanism design   67
mechanisms, properties   106
metal forming   12
   properties   105
metamorphosis   8
mild steel   12, 14
mock-up 35
modelling   22, 42, 108
monster design   12, 14
mood board   66
mordants   74
moving display project   64
musical instrument project   32
   tuning   37
   sounds   34

needs analysis   40
   customer's   21
   identifying   90

one-off production   120
op art   10
optical illusions   6, 8
ornament   15
ornamental steelwork   15

packaging, food   82
paper weight   15
pasta production   26, 29
   project   24
pattern, textile   61
perception   9
pilot plant   83
plastic properties   105
pneumatics, properties   106
pockets, textile   62
portion size   98
presentation   111
pricing   100
printing, fabric   77
process control   30
product evaluation   80
   testing   116

production   104, 120
   quality   125
profiling clients   20
project, ceramics design   44
   colour design   70
   fish dish   44
   flashing lights   56
   Italian food   24
   layered dessert   78
   monster   12
   moving display   64
   musical instrument   32
   promotional items   84
   sheet steel   18
   torch   38
   wind generator   50
   window display   64
promotional items project   85
   products   86
properties of materials   105
prototyping   118

quality assurance   30
   control   29
   production   123, 125

rolling clay   48
rotor blades   52
rust   17

safety design   99
scoring (cutting)   14
seams, textile   61
setting, food   107
shaping steel   16
sheet material   20
shelf-life   98
Shibori dyeing   76
sketches, thumbnail   22
sound generation   34
specification, design   21, 57, 92
specification, dessert   81
spider diagram   86
star diagram   82
steel properties   14
   bending   14
   cutting   16
   finishes   17
   folding   18

joining   18
rust   17
sheet   18
steelwork   15
structures   54
surface finish   17
sweetening, food   107

teamwork   109
teenage market   20
templates   47, 61
tension   14
tessellations   11
testing products   116
textile buttonholes   62
   colouring   72
   decoration   63
   design   96
   fastenings   62
   finishing 61
   hems   61
   joining   61
   pattern   61
   pockets   62
   reinforcement   62
   seams   61
textiles   58, 60
thickening, food   106
thinning, food   107
thumbnail sketches   22
tie and dye   75
torch project   38
tritik dyeing   75
tuning, musical instrument   37

user needs   41

vacuum forming   83
visual communication   66
volume production   29, 120

warning beacons   56
weight control   31
wind generator project 50
windmills   52
window display project   64
wood properties   105
work planning   13